ISDN Explained

Second Edition

ISDN Explained

Worldwide Network and Applications Technology

Second Edition

John M. Griffiths
BT, UK

with

Peter Adams
John Hovell
Kevin Woollard
Derek Rumsey
David Davies
John Atkins
Paul Challener
Malcolm Jones
Graham Hudson
Geoff Morrison
Paul McDonald
Richard Boulter

JOHN WILEY & SONS
Chichester • New York • Brisbane • Toronto • Singapore

Reprinted February, August 1994, March 1995

BT will donate the royalties from the sale of this book to charity

Other Wiley Editorial Offices

John Wiley & Sons, Inc., 605 Third Avenue,
New York, NY 10158-0012, USA

Jacaranda Wiley Ltd, G.P.O. Box 859, Brisbane,
Queensland 4001, Australia

John Wiley & Sons (Canada) Ltd, 22 Worcester Road,
Rexdale, Ontario M9W 1L1, Canada

John Wiley & Sons (SEA) Pte Ltd, 37 Jalan Pemimpin #05-04,
Block B, Union Industrial Building, Singapore 2057

Library of Congress Cataloging-in-Publication Data

Griffiths, J. M. (John M.)
ISDN explained : worldwide network and applications technology /
John Griffiths with Peter Adams . . . [et al.]. — 2nd ed.
p. cm.
Includes bibliographical references and index.
ISBN 0 471 93480 1
1. Integrated services digital networks. I. Adams, P. F. (Peter
F.) II. Title
TK5103.7.G75 1992
621.382—dc20 91-7466
CIP

British Library Cataloguing in Publication Data

A catalogue record for this book is available from the British Library

ISBN 0 471 93480 1

Body text typeset in 10/12pt Palatino by Acorn Bookwork, Salisbury, Wiltshire
Printed and bound in Great Britain by Bookcraft (Bath) Ltd

Contents

Preface

Five years ago, when the first edition of this book was written, ISDN was a matter of specialist interest; now it is becoming commonplace and for that reason is routinely part of telecommunications courses at all levels. It is therefore right that this book should be issued in a form suitable for student use.

This book is pitched at a level deeper than the general "gee-whiz, isn't it wonderful" by going into the principles of all facets of the ISDN operation. It does not pretend to go down to the detail needed to build network or terminal equipment. The full details would require a book several times the size of this but it does set out the principles on which the whole lot works. This book gives an overview to set the operations in context, to introduce Standards, Recommendations and operating manuals.

Changes to this book from the second edition have involved the rewriting of some areas in which I felt there was scope for making explanations easier to understand; in this connection I would like to thank Iain Richardson of The Robert Gordon University for the graphical way of explaining the DCT in chapter 8. I have also included a "route map" to ISDN, printed on the inside cover, which sets out to clarify the ISDN story by indicating the relationship between the various strands. Readers may also wish to note that references to "CCITT" are now more properly to "ITU-T" as a result of changes in that organisation.

The work recorded in this book is the result of many people's inspiration. I would like to thank the contributors to this book who have filled in their own areas of expertise. However, in addition to all those mentioned by name, there were also many others whose advice and information have also contributed and whom I must thank. I would also like to compliment Peter Adams, who wrote chapter 4, for his leadership of the team at BT Labs whose work on local network transmission and the 2B1Q line code has led to what has become the standard local network transmission system. Finally I would like to thank Sue Hornsby for her patience in typing and retyping this book.

John Griffiths
British Telecom Laboratories
June 1994

Note to lecturers:

To assist in the organisation of courses on this topic, a PC floppy disk is available with more than 60 of the illustrations from this book in Microsoft® PowerPoint® format which can be used for OHP slides. The disk also includes an executable file which calculates the loss probability for traffic consisting of any mix of single and multi-slot traffic which enables you to extend the calculations set out on page 182 to more complex situations. If you are interested in finding out more about the disk, please contact the Text Books department, John Wiley & Sons Ltd., Baffins Lane, Chichester, Sussex PO19 1UD. Tel. (01243) 779777.

Biographies

John Griffiths (who wrote all the unattributed sections and edited the book) was educated at Ealing Grammar School and the University of Manchester whence he graduated with a 1st in Electrical Engineering. Out of perversity he then spent a year on flying training in the Royal Air Force before joining British Telecom (then the Post Office) in 1967, working on early PCM systems. In 1971 he was promoted and bypassed offers of jobs in the proper subjects of switching and transmission, to take a job in the wilderness (as it was then) of Local Telecommunications. From early work on CATV and Local Network Transmission, this evolved to ISDN and its protocols and has progressed to Broadband ISDN. He is now Division Manager of the Digital Services Division at British Telecom Laboratories, Martlesham, and a Fellow of the Institution of Electrical Engineers. On behalf of that organization he is the UK representative on the Council of the International Symposium on Subscriber Loops and Services (ISSLS). He is married with two children and for hobbies he is a flying instructor and train spotter.

Peter Adams (who wrote Chapter 4) joined British Telecom in 1966 as a student apprentice. After obtaining a BSc in electronic engineering from Southampton University in 1970, he joined British Telecom Research Laboratories where he worked on telecommunications applications of digital signal processing. Since 1979 he has headed a group concerned with speech-band data modems and subsequently local network transmission and maintenance. Currently he leads a Local Network Exploitation section.

John Hovell (Section 5.1) joined British Telecom as an apprentice in 1967. He moved to the Research Department in 1974, after graduating with an honours degree in electrical and electronic engineering, where he worked on line card design for early digital exchanges. In 1981 he obtained an MSc in telecommunications systems and went on to lead a group responsible for the ISDN user/network interface. Recently this work has expanded to include testing, in particular the application of Formal Methods to the proving of signalling protocols. He is a chartered engineer and member of the IEE.

Kevin Woollard (Section 5.2 and Appendix A) graduated with a BSc degree in physics from Bath University in 1983. He returned to British Telecom to work on network related aspects of ISDN and in particular has implemented a Layer 2 handler for an experimental ISDN exchange. He was involved in ISDN conformance test specifications as a member of the European group responsible for the WAN-CTS conformance tests for ISDN equipment, working on aspects of automatic test case generation from protocol specifications.

Derek Rumsey (Section 5.3) joined British Telecom in 1965. Following a period working on the development of microwave radio links, he joined a division responsible for the system design of data networks where he specialized in packet network protocol. In 1981 he became involved in the development of protocols for ISDN access and currently heads a group responsible for the specification of signalling systems for access to British Telecom's ISDN.

David Davies (Section 5.4) joined British Telecom in 1974 after graduating from Oxford University. He became involved in the development of protocols for the ISDN in 1982 and now heads a group responsible for international Standards relating to the ISDN customer access. Until 1990, he chaired the ETSI committee which is responsible for developing the ISDN user–network protocol Standards and is a delegate to the CCITT Study Group which studies this subject.

John Atkins (Chapter 7) is Head of the Data Communication Networks Section at British Telecom Laboratories. His background includes R&D in digital switching systems for voice and data services, exploiting both circuit and packet modes of operation. His present interests lie mainly in the networking of non-voice services, and span LANs, MANs, and public wide area networks, with special reference to the ISDN.

Paul Challener (Section 8.1) gained a 1st class honours degree in electronic engineering from Nottingham University in 1974 and became a Chartered Engineer in 1981. He is well known internationally in the speech coding community and has contributed extensively in IEE and IEEE activities both at home and abroad. He is currently Head of the Voice Terminals Section at British Telecom Laboratories, concentrating on developments involving interactive speech and high quality audio processing. He has worked on high quality speech and music coding since 1982.

Malcolm Jones (Section 8.3) joined the Research Department of the Post Office in 1963 and worked on communications systems for government departments until 1970 when more specialist activities on maximizing the capacity of coaxial cables for submarine and land-line applications were undertaken. This was followed by data modem design using adaptive equalizers for higher speed operation over the PSTN. This led to general terminal development and international standardization work for the PSTN, particularly in the areas of facsimile and teletex. Current activities are still in the area of terminal development including for the ISDN in the form of PC based 64-kbit/s terminal products and for satellite based networks with their unique broadcasting ability.

Graham Hudson (Section 8.4) began his career in British Telecom as a technician apprentice in 1966, and then, with a BT scholarship, he obtained an honours degree in electrical and electronic engineering at City University, London. In the early 1980s, as Head of the Advanced Videotex Terminals Group, he was responsible for the development of ISDN photovideotex terminals. During this time he was chairman of the ESPRIT PICA project and the ISO JPEG, both concerned with the development and selection of photographic coding techniques. He is now Head of the Home and Office System Section at British Telecom Laboratories. He is a chartered engineer and a member of the IEE.

Geoff Morrison (Section 8.5) joined the Visual Telecommunications Division of the British Post Office Research Laboratories after graduating from St John's College, Cambridge in 1971. He has worked on video terminal and transmission aspects, both analogue and digital, mainly for videotelephone and videoteleconferencing services. He is head of a group responsible for low bit-rate video coding Standards and was a UK representative in the CCITT Specialists Group which formulated Recommendation H.261. His current interests include coding for storage devices and for asynchronous transfer mode networks.

Paul McDonald (Section 9.4) joined British Telecom in 1987 after graduating with a BEng (Hons) in Electronic and Microprocessor Engineering from Strathclyde University. Since then his work has encompassed research into a broad range of ISDN topics including the ISDN basic rate interface, broadband ISDN, ATM techniques, and SDH transmission Standards. Over the past few years he has been investigating the problems of interworking mobile systems with the ISDN, in particular the effects on access and CCITT No 7 signalling protocols.

Richard Boulter (Section 9.5) joined British Telecom in 1963 as a student apprentice. After graduating from Birmingham University in 1967 he joined a data transmission research group at the Research Station. In 1974, he was responsible for specifying the synchronization equipment for the UK digital network. After further work on line-card developments for digital exchanges and a six-month consultancy period with SHAPE Technical Centre, he headed a section concerned with systems evolution. Since then, he has been involved in specification work on BT's ISDN, studies into the evolution of the local network and acted as manager of two RACE collaborative projects on broadband ISDN.

Chapter 1

Network Evolution

1.1 THE TELEGRAPH

The earliest telecommunications systems were optical. Although the use of beacon fires to warn of anticipated dangers is widely known, it is difficult to regard as telecommunication a system which only allowed the transmission of one bit of information over a period of several weeks. There were several attempts at more flexible systems using large shutters or movable arms which could be mounted on tops of hills (Figure 1.1). These were observed from adjacent hills and relayed on. The system which came into virtual universal use was the French semaphore system invented in 1792. By 1830 much of Europe was covered by such networks.

In 1837 the first practical electrical telegraph appeared. In Europe Cooke and Wheatstone produced their five-wire telegraph (Figure 1.2). In the same year Samuel Morse invented his well-known system, although it was another seven years before it was put into service. For the next 40 years telegraphy evolved technically and in its services. The five wires of the Wheatstone system were complex and expensive but the ability of unskilled operators to use it made it in some situations more attractive than the morse system. The ideal was a system which only used two wires (or one and an earth return) and was simple to operate. The eventual result of this was the teletypewriter developed in 1915 by Kleinschmidt and Morkrum in the USA.

In functional terms by 1875 the telegraph network was simple with European countries each offering a state monopoly telegram service. In the USA, while not an actual monopoly, the telegraph network was dominated by Western Union. Hence a resident of that time could have considered that he or she had an Integrated Services Digital Network (ISDN), there only being one network for all the services available at that time and this being digital in nature.

1.2 THE TELEPHONE ARRIVES

The coming of the telephone in 1875 removed both the integration and digitalization of the network. Bell's early attempts at telephony involved

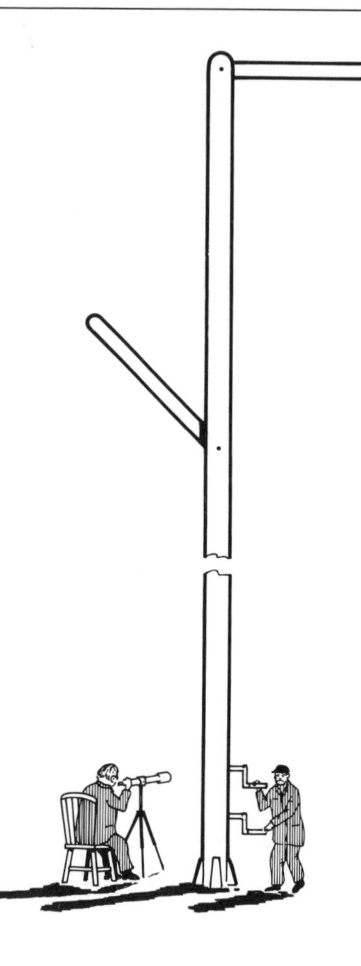

Figure 1.1
The semaphore.

operating a make–break contact by a microphone diaphragm in a binary way, but his early success relied on the change of magnetic circuit impedance and hence magnetic field in a coil due to movements of a diaphragm. However, this lacked sensitivity. Elisha Gray made real progress when he made the contact softer using a diaphragm connected to a wire immersed in a fluid giving a resistance change proportional to the audio pressure change (Figure 1.3). Edison's carbon microphone invented in 1878 was a more convenient way of achieving the same effect.

The scene was therefore set for two clearly separate telecommunications networks. On the one hand, the telephony network carrying signals whose magnitude was fairly linearly related to the sound pressure incident upon the sending microphone and were hence called 'analogue'. Its big advantage was that it could be used by unskilled operatives; communication simply involved speaking and listening and what can be more natural than

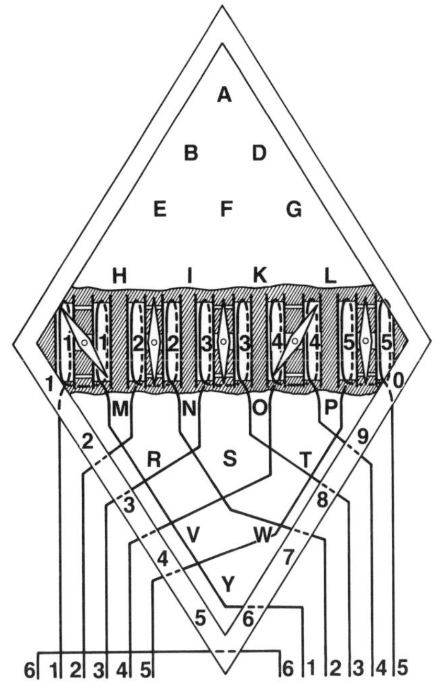

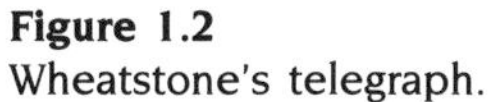
Figure 1.2
Wheatstone's telegraph.

that? On the other hand, the digital telegraph network had the advantage that a hard copy of the communication is produced. In technical terms the telegraph network had further advantages; at a very early stage a means of relaying (Figure 1.4), or repeating, signals was developed such that operation over long distances was possible. This has always presented a problem for telephony and was initially solved by using wires of extraordinary

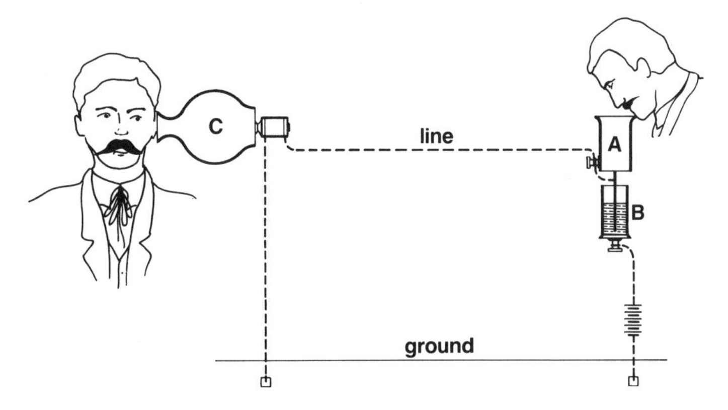

Figure 1.3
Elisha Gray. Instruments for transmitting and receiving vocal sounds telegraphically. Caveat filed 14 February 1876.

Figure 1.4
Telegraph repeater 1901.

thickness. The invention of the triode by Lee de Forest in 1906 meant that attenuation of telephony signal could be compensated for, but any noise or non-linear distortion once acquired by the signal was there for good. Even the use of amplification presented problems in that, if it was not carefully controlled, systems could become unstable and oscillate. In a telegraph network one merely had to identify whether a signal was present or not.

Thus the science of telephony became one of balancing the economies to be gained by concentrating traffic into a small number of large switching centres, against the need to maintain an acceptable speech quality by having switching centres close to the customer and hence numerous.

The convenience of the telephone meant that its numbers grew very rapidly and, because of the pressure of demand, technical problems were at least made manageable if not overcome. For transmission purposes the most fundamental change was to make the trunk network into a four-wire system so that the two directions of transmission were separated and could be amplified without the compromises required for full duplex operation on one pair of wires. The other innovation in the first decade of the twentieth century was the introduction of automatic switching. Although this increased the complexity of operations required of the user, it meant that the person power and accommodation required to provide switching could be kept within reasonable bounds in an ever-expanding network.

1.3 TELEX

The expansion of the telephone network was such that in the early 1900s the public telegraph ceased to be the dominant form of telecommunication. On the other hand it provided an essential service to those without access to telephones or where a written record was important. The need for the written record and the development of the page teleprinter in 1928 led to the evolution of the telex service in which the telegraph was extended to customers' premises. Initially this was offered over the existing telephone network using what would now be called modems which, at the customer premises, converted the d.c. teleprinter signals into modulation of a 1500 Hz signal for transmission across the telephone network. In practice this presented problems as connections which were entirely satisfactory for telephony purposes often gave an excessive error rate for telex purposes. For this reason all countries eventually followed the lead of Germany, which had in 1936 set up a separate switched telex network directly carrying the d.c. teleprinter signals. An additional advantage is that between the switching nodes 24 telex channels could be carried on one 4 kHz telephony bearer. The narrow bandwidth requirement meant that it was easier to extend telex networks overseas. Even though the telex network was intended to create a simultaneous two-way connection, it was also relatively easy to interface it to a store and forward facility where traffic or (in the case of HF radio) circuit conditions dictated. This further extended the attraction of telex to the business community.

Thus by the 1950s in most countries there were three major types of public networks:

(a) The public telegraph network, by then in decline but with 30 years still to go.

(b) The telephony network, by far the most popular owing to its ease of use, based on a nominal 4 kHz bandwidth and using analogue techniques, which in the customer's access link had not changed in transmission principle since 1880, and in signalling principle since about 1910.

(c) The 50 baud telex network, proving indispensable to the commercial world because of its provision of a written record and the ability to work unattended.

In the trunk network telephone signals were frequency division multiplexed to carry 12 channels (a 'Group') on a single pair in the range 60–108 kHz. These groups were then assembled into supergroups of five Groups for transmission on coaxial cables. These supergroups could then be further assembled into master groups. Switching was by electromechanical switches, either a developed form of Strowger's original step-by-step switch of 1898 or motor driven switches. The crossbar switch of 1938 is a more refined way of connecting lines and its principles (but not its mechanism) carry through to the present day. Intercontinental connections were largely by HF radio. In 1956 North America and the UK were first linked by telephone cable, and in 1965 by geostationary satellite. In the 1960s there was also a general extension of the use of microwave links as an alternative to coaxial cable and the sealed reed relay also became an attractive alternative to the open contact of previous generations of switch.

1.4 THE COMPUTER AND THE TRANSISTOR

In the 1950s the next two significant players came on the scene. First the concept of the electronic computer; at this stage it was really only the concept which was significant, because the implementation using vacuum tubes, while an advance on previous electromechanical techniques, would not survive long. However, the concept of being able to store and manipulate large amounts of data would surely also lead to the need to transfer a lot of data from place to place.

Secondly the invention of the transistor in 1948 meant that computers of then unimagined storage capacity and processing power could be realized in a small size and consuming little power. These would be able to absorb enormous amounts of data communications capacity. The computer could also be used to control telephone exchanges.

Although telex networks might have been the obvious candidates for upgrading to higher data rates for transferring computer data, their roots lay in electromechanical technology and very little was done in this direction; instead the telephone network was pressed into service. As in the early telex days, modems were designed which could take the baseband data signals and modulate them upon audio carriers to be transmitted across the telephony network. The first modems were available in the early 1960s and operated at 100 baud. Owing to improved modulation schemes, and improvements to the telephone network itself, the error performance of the modem was acceptable and the problems of the early telex modems did not recur in general although particular circuits gave difficulty. Over the years modem performance has been steadily improved until by the end of the 1980s, 9.6 kbit/s can be carried over a typical switched telephony circuit. However, this progress cannot go on indefinitely. Shannon's famous theorem gives an upper bound to the information capacity of a channel, based on its bandwidth and signal-to-noise ratio.

$$I = F \log_2 (1 + S/N)$$

where F is the bandwidth available and S/N is the signal-to-noise ratio. For a telephony channel the useful bandwidth is about 3 kHz and the signal-to-noise ratio may be about 30 dB. Hence the capacity is of the order of

30 kbit/s. Experience has shown that to get within a factor of 2 or 3 of Shannon's limit is a notable achievement and little more can be expected in that direction.

1.5 DATA NETWORKS

Even at these data rates the telephony network leaves much to be desired for intercomputer communications. In particular call set-up on electromechanical systems takes several seconds, and with computer processors operating at millions of operations a second, the time to set up a call is often embarrassing, even when a processor is able to delegate such tasks to peripheral buffers. For this reason in the 1970s consideration was given to further specialized networks for intercomputer communication. These fell into two categories. The traditional solution was a circuit-switched network similar in principle to the telex network, but able to operate at kbit/s rather than bit/s, and with fast call set-up. The other solution was more revolutionary and considered more deeply the needs of the computer community. The concept could be compared with the postal service in which a packet is dropped into a mailbox. This packet could be a postcard or a small parcel and is routed to its destination on the basis of the address on the outside. In telecommunications packet networks, data is preceded by a header containing an address by which the packet is routed to its destination. This is discussed in more detail in Chapter 7. Although this is a very flexible solution, it does require a lot of high speed intelligence applied to each packet at each switching node which has its cost and throughput penalties. The consensus is that where capacity is of greatest consequence then circuit-switched networks are best. Where flexibility is important packet-switched networks are best.

Both packet- and circuit-switched networks have been implemented. Early examples of packet networks are Spain (Iberpac), UK (PSS), France (Transpac), Germany (Datex P), Switzerland (Telepac), USA (GTE Telenet, Tymnet) and Canada (Teleglobe). Circuit-switched networks were established in the Nordic countries (Norway, Sweden, Denmark and Finland) and the Federal Republic of Germany.

Most developed countries now have such data networks. In the case of circuit-switched services they will offer data rates typically of 9.6 kbit/s, with call set-up times of a few hundred milliseconds. In the case of packet networks, connections may be made at up to 48 kbit/s but throughput is limited by the flow control mechanisms in the network, depending on network or destination capacity.

By the end of the 1980s the telegraph network had virtually ceased to exist. In some countries the telegram service was no more; in others it made use of the telex network. Telex networks were booming but probably at their peak, coming under pressure from facsimile services operating over the telephone network and also (in some countries) teletex services operating on public data networks. The public switched data networks were expanding but with competition from the omnipresence of the telephone network and its modems.

1.6 DIGITAL TELEPHONY

For the next stage in the proceedings one must go back to 1937 when Alec Reeves conceived pulse code modulation (PCM) of speech signals, specifically to overcome the aforementioned problem of the accumulation of distortion and noise in analogue networks. The process is given in detail in the next chapter; in outline it consists of measuring the analogue speech signals 8000 times a second, encoding them into one of 256 levels represented by 8 binary bits, and hence converting the analogue speech into a binary stream at 64 kbit/s. The advantage of this is that if the signal is regenerated before noise is sufficient to make the two binary states difficult to distinguish, then there is no accumulation of distortion. When first invented no devices were available which could make PCM economic. Wide-scale use of PCM for telephony transmission did not occur until the late 1960s. The individual 64 kbit/s streams were multiplexed by interleaving them to carry 24 (North America) or 30 (Europe) channels on pair-type cables. Primary rate multiplexes could then be further interleaved to carry more channels on coaxial cables or microwave radio systems.

In the 1970s it was recognized that Very Large Scale Integrated Circuits (VLSI) and optical fibre transmission would make PCM the obvious choice for future telephony networks, for transmission and switching purposes. The developments of the 1970s came to fruition in the 1980s, with all major network operators replacing their networks and switching nodes by digital facilities. The switches were entirely electronic with computers (usually called processors) to control them, and interprocessor links (called common channel signalling) to control call set-up. Note, however, that the digital PCM operation does not extend to include the local access loop to the customer, which remained the analogue two-wire link with a telephone whose transmission antecedents go back to Edison, with a pulse signalling system introduced in about 1910 or possibly a multifrequency keypad signalling system first used in the USA in 1960. Nevertheless, conversion of the telephony main network to digital has enormous implications. It is now an infrastructure of universal extent which not only can carry telephony, but can carry data at 64 kbit/s. As data and telephony can now be integrated on the one network it is called an Integrated Digital Network or IDN.

1.7 THE ISDN

The next stage is to take the digital channels all the way to the customer so that all services can be integrated on one bearer. This is the Integrated Services Digital Network or ISDN. CCITT defined the ISDN as:

A network evolved from the telephony IDN that provides end-to-end digital connectivity to support a wide range of services, including voice and non-voice services, to which users have access by a limited set of standard multipurpose customer interfaces.

The terminology 'Integrated Services (digital) network' was invented in 1971 by CCITT Study Group XI; the parentheses were soon abandoned. Two challenges faced the ISDN proponents.

1. To develop a system which would carry the 64 kbit/s digital channels to the customer. The local network mainly consists of copper pairs to provide the final link to the customer which are designed primarily to carry signals up to 4 kHz. The 4 kHz limit is not absolute (except in some North American situations where loading coils were inserted), but attenuation of the pair and crosstalk from other pairs increases as frequency increases. One solution is to substitute the copper pair by an optical fibre. However, even though fibres are comparable in price to copper pairs, the cost involved in the changeover is prohibitive unless the demand is very significant and this is difficult to substantiate in early days. Copper pairs also have the advantage of being able to carry power for a telephony terminal, which is desirable for reliability and emergency situations. Chapter 4 considers the use of copper pairs as the ISDN bearer.
2. The interface to the customer should be appropriate for a multiplicity of services. Data services are normally terminated in a multipin socket into which a single device can be plugged. Several telephones may be connected in parallel across the incoming analogue telephone line. What is needed is a system which could allow several terminals of different types (e.g. VDU, telephone, personal computer, facsimile machine) to be simply connected to the network, and a signalling system which allowed the terminals to be appropriately called and make calls. It was also decided that with this multiplicity of services it was unreasonable to restrict customers to the use of only one channel at a

time. For this reason a basic access offering two 64 kbit/s channels is standardized upon, allowing the use of two terminals at a time. These channels are called 'B' channels. A 16 kbit/s signalling channel is also offered, which may also be used to provide access to a packet-switched service. This channel is called a 'D' channel. The interface identified is a four-wire bus structure to which up to eight terminals can be connected in parallel, whose operation is described in detail in Chapter 5.

I am often asked how the terminology 'B' and 'D' arose. In early discussions the Analogue channel was referred to as an 'A' channel and hence the next letter of the alphabet was chosen to represent an ISDN channel particularly as this could be construed as 'B' for 'binary' or 'bits'. The signalling channel was regarded as a small increment on the traffic channel and hence, borrowing from calculus, was known as the delta channel, represented by the Greek letter 'Δ'. Unfortunately few typewriters could generate this character and so it became the custom to use the Roman equivalent 'D'.

The ISDN is a complex mix of network capability and customer premises equipment supporting applications which can exploit the capability. It is the standard exchange line of the future.

1.8 EARLY ISDNs

Although discussions on the ISDN had been in progress throughout the 1970s, it was not until 1984 that CCITT published Recommendations for the Standards to be adopted for ISDN services and interfaces. Prior to that several administrations were in a position where they were installing an extensive network of digital exchanges which could be used for ISDN service. Rather than delay the exploitation of the network they introduced services using proprietary Standards.

The first service was introduced in June 1985 by British Telecom. Marketed as IDA (Integrated Digital Access) pilot service, it was initially based on one exchange in London. By the end of 1985 it was extended to four exchanges (two in London and one in Birmingham and Manchester) with remote multiplexers as shown in Figure 1.5. By 1988 it was available in 60 towns throughout Britain. It offers two simultaneous traffic channels: one at 64 kbit/s and the other at 8 kbit/s. The original terminal apparatus included digital telephones with data ports and a termination which could offer a whole series of standard data ports. However, because of the requirements of the regulatory regime in the UK, only a standard X.21 interface was offered from 1986.

Figure 1.5
Pilot ISDN coverage in the UK 1986.

Another early pre-standard ISDN service was project Victoria, offered by Pacific Bell in 1986 in Danville, California. This offers to a customer five RS232C ports which can be configured to operate at various rates from 9.6 kbit/s down to 50 bit/s, together with an analogue telephone. The whole is multiplexed on to an 80 kbit/s line rate.

In France since 1985 the Transcom service offers 64 kbit/s interconnect over the digital telephony network using X.21 and V.35 interfaces to the customer. Also in 1986 in the USA the local 'Public Switched Digital Service', combined with the nationwide Accunet 56 service, has offered a similar service, but restricted to 56 kbit/s owing to the continued presence of circuits from which bits are stolen for signalling purposes. This will be mentioned again in Section 3.1. A common feature of all these systems was the restriction of the local network bit-rate to around 80 kbit/s. The higher bit-rate required for two 64 kbit/s channels was only agreed after extensive testing of the local networks of many member countries of CCITT. Shakespeare summed it up with '2B or not 2B, that is the question'.

The first CCITT ISDN Recommendations formally appeared in 1984 in their 'Red Book' and systems more or less to these Standards have been put into service in Oak Brook, Illinois, USA at the end of 1986, in Mannheim and Stuttgart, Germany in 1987, and also in Brittany, France in 1987. By 1988 about forty different trials or services to Red Book Standards were in progress.

However the next generation of CCITT Recommendations issued in their 1988 'Blue Book' do contain significant changes and clarifications which will lead to truly international Standards.

1.9 INTERNATIONAL STANDARDS

The cost of developing modern equipment is very high. The design and verification of a single VLSI chip costs millions of pounds or dollars; however, once designed, the unit production costs are low. For software the situation is even more extreme; development costs are very high but replication costs are almost zero. For this reason it is important that any product gets the widest possible use to spread development costs. Worldwide Standards against which all telecommunications end-users may purchase are therefore very important. This also eases interworking between different networks. In telecoms the main Standards are developed and issued by the CCITT (the French abbreviation for the International Telephone and Telegraph Consultative Committee). The Recommendations are issued in several series. These are the ones which are of greatest relevance to ISDN:

V series for data/communications over the switched telephone network.

X series for dedicated data communications networks.

T series for terminal equipment and protocols for telematic services.

Q series for signalling systems.

G series for telecommunication transmission systems.

E series for operations, numbering and routing.

H series for line transmission of non-telephone signals.

I series for general ISDN matters.

Most of this book will refer to the I series of Recommendations and these are listed in Table 1.1. Where there is an alternative number in brackets this means that the I series Recommendation refers out to a recommendation in another series. A particular feature of the ISDN is the establishment of a reference configuration for the ISDN connection between the customer and the exchange. This is shown in diagrammatic form in Figure 1.6. TE1 is a terminal which complies with ISDN user–network interface Recommendations such as a digital telephone, data terminal equipment or an integrated workstation. TE2 is also a terminal which complies with some other interface Standards (e.g. V or X series) at reference point R, and hence

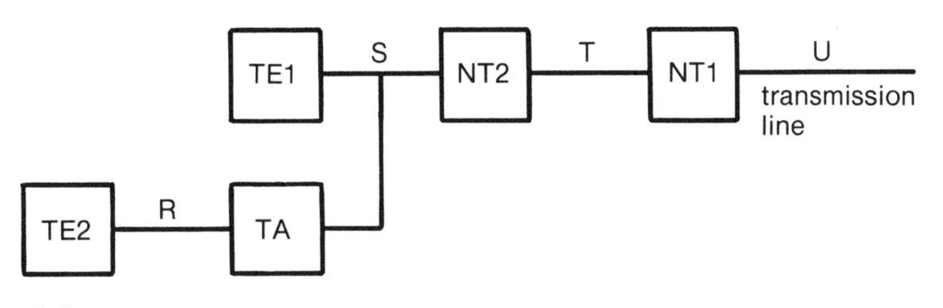

Figure 1.6
ISDN reference configuration.

Table 1.1
CCITT Recommendations on ISDN.

Rec. No	
I.110	Preamble and general structure of the I series Recommendation for the integrated services digital network (ISDN).
I.111	Relationship with other Recommendations relevant to ISDNs.
I.112	Vocabulary of terms for ISDNs.
I.113	Vocabulary of terms for broadband aspects of ISDN.
I.120	Integrated services digital networks (ISDN).
I.121	Broadband aspects of ISDN.
I.122	Framework for providing frame mode bearer services.
I.130	Method for the characterization of telecommunication services supported by an ISDN and network capabilities of an ISDN.
I.140	Attribute technique for the characterization of telecommunication services supported by an ISDN and network capabilities of an ISDN.
I.141	ISDN network charging capabilities attributes.
I.150	B-ISDN asynchronous transfer mode functional characteristics.
I.200	Guidance to the I.200 series of Recommendations.
I.210	Principles of telecommunication services supported by an ISDN and the means to describe them.
I.211	B-ISDN service aspects.
I.220	Common dynamic description of basic telecommunication services.
I.221	Common specific characteristics of services.
I.230	Definition of bearer service categories.
I.231	Circuit-mode bearer service categories.
I.232	Frame-mode bearer service categories.
I.233.1	ISDN frame mode bearer services (FMBS)—ISDN frame relaying bearer service.
I.233.2	ISDN frame mode bearer services (FMBS)—ISDN frame switching bearer service.
I.240	Definition of teleservices.
I.241	Teleservices supported by an ISDN.
I.250	Definition of supplementary services.
I.251	Number identification supplementary services.
I.252	Call offering supplementary services.
I.253	Call completion supplementary services.
I.253.1	Call waiting (CW) supplementary service.

Table 1.1 (*continued*)

Rec. No	
I.254	Multiparty supplementary services.
I.255	Community of interest supplementary services.
I.255.3	Multi-level precedence and preemption service (MLPP)—Description preference service.
I.255.4	Priority service.
I.256	Charging supplementary services.
I.257	Additional information transfer.
I.310	ISDN—Network functional principles.
I.311	B-ISDN general network aspects.
I.312(Q.1201)	Principles of intelligent network architecture.
I.320	ISDN protocol reference model.
I.321	B-ISDN protocol reference model and its application.
I.324	ISDN network architecture.
I.325	Reference configurations for ISDN connection types.
I.326	Reference configuration for relative network resource requirements.
I.327	B-ISDN functional architecture.
I.328(Q.1202)	Intelligent network—service plane architecture.
I.329(Q.1203)	Intelligent network—global function plane architecture.
I.330	ISDN numbering and addressing principles.
I.331(E.164)	Numbering plan for the ISDN area.
I.332	Numbering principles for interworking between ISDNs and dedicated networks with different numbering plans.
I.333	Terminal selection in ISDN.
I.334	Principles relating ISDN numbers/subaddresses to the OSI reference model network layer addresses.
I.335	ISDN routing principles.
I.340	ISDN connection types.
I.350	General aspects of quality of service and network performance in digital networks, including ISDN.
I.351(G.821/2)	Recommendations in other series concerning network performance objectives that apply at reference point T of an ISDN.
I.352	Network performance objectives for connection processing delays in an ISDN.
I.361	B-ISDN ATM layer specification.
I.362	B-ISDN ATM adaptation layer (AAL) functional description.
I.363	B-ISDN ATM adaptation layer (AAL) specification.
I.370	Congestion management for the ISDN Frame Relaying Bearer Service.
I.410	General aspects and principles relating to Recommendations on ISDN user–network interfaces.
I.411	ISDN user–network interfaces—reference configurations.
I.412	ISDN user–network interface—interface structures and access capabilities.
I.413	B-ISDN user–network interface.
I.420	Basic user–network interface.
I.421	Primary rate user–network interface.

Table 1.1 (*continued*)

Rec. No	
I.430	Basic user–network interface—Layer 1 specification.
I.431	Primary rate user–network interface—Layer 1 specification.
I.432	B-ISDN user–network interface—physical layer specification.
I.440(Q.920)	ISDN user–network interface data link layer—general aspects.
I.441(Q.921)	ISDN user–network interface, data link layer specification.
I.450(Q.930)	ISDN user–network interface Layer 3—general aspects.
I.451(Q.931)	ISDN user–network interface Layer 3 specification for basic call control.
I.452(Q.932)	Generic procedures for the control of ISDN supplementary services.
I.460	Multiplexing, rate adaption and support of existing interfaces.
I.461(X.30)	Support of X.21, X.21 bis and X.20 bis based data terminal equipments (DTEs) by an integrated services digital network (ISDN).
I.462(X.31)	Support of packet mode terminal equipment by an ISDN.
I.463(V.110)	Support of data terminal equipments (DTEs) with V series type interface by an integrated services digital network (ISDN).
I.464	Multiplexing, rate adaption and support of existing interfaces for restricted 64 kbit/s transfer capability.
I.465(V.120)	Support by an ISDN of data terminal equipment with V series type interfaces with provision for statistical multiplexing.
I.470	Relationship of terminal functions to ISDN.
I.500	General structure of the ISDN interworking Recommendations.
I.510	Definitions and general principles for ISDN interworking.
I.511	ISDN-to-ISDN Layer 1 internetwork interface.
I.515	Parameter exchange for ISDN interworking.
I.520	General arrangements for network interworking between ISDNs.
I.530	Network interworking between an ISDN and a public switched telephone network (PSTN).
I.540(X.321)	General arrangements for interworking between circuit-switched public data networks (CSPDNs) and integrated services digital networks (ISDNs) for the provision of data transmission.
I.550(X.325)	General arrangements for interworking between pocket-switched public data networks (CSPDNs) and integrated services digital networks (ISDNs) for the provision of data transmission.
I.560(U.202)	Requirements to be met in providing the telex service within the ISDN.
I.601	General maintenance principles of ISDN subscriber access and subscriber installation.
I.602	Application of maintenance principles to ISDN subscriber installations.
I.603	Application of maintenance principles to ISDN basic accesses.
I.604	Application of maintenance principles to ISDN primary rate accesses.
I.605	Application of maintenance principles to static multiplexed ISDN basic accesses.
I.610	OAM principles of the B-ISDN access.

needs a terminal adapter (TA) to adapt to the ISDN user–network interface. NT2 is some sort of customer premises switching function such as a PABX or LAN, or it may be a null function if it is not required. Reference points S and T apply to physical interfaces which are defined in the I series of Recommendations. NT1 is on the customers premises and provides the proper physical and electromagnetic termination of the network. The transmission line (or U reference point) is terminated by the NT1. Chapter 4 considers the interface at the U reference point and Chapters 5 and 6 consider the interface at the S and T reference point.

1.10 OPEN SYSTEMS INTERCONNECT (OSI); THE 7-LAYER MODEL

OSI is a term for the agreed International Standards by which systems should communicate. In telecommunications, as in most other activities, it takes teamwork to provide and utilize the facilities offered. Most basically transmission media such as optical fibres or copper pairs must be provided, in addition signalling and speech coding must also be provided, and all this will be wasted if there are no customers who wish to talk to one another. Thus there is a natural layering of the telecommunications process. The International Standards Organization (ISO) has formalized this into seven layers for the interworking of computers, terminals and applications.

In the 7-layer model it is assumed that one has a physical connection such as an optical fibre, copper pair or coaxial line. Upon this is built:

Layer 1 — the physical layer defines the characteristics of the signal to be transferred over the bearer. It covers such things as pulse amplitudes, line coding, transmission rates, connectors, and anything else needed to transfer digits satisfactorily.

Layer 2 — the link layer provides discipline for the assembling of the digits. It provides error detection and correction by assembling the digits into frames. At present all Layer 2 formats are derived from a standard known as High Level Data Link Control (HDLC).

Layer 3 — the network layer ensures that messages are routed to the appropriate destinations, and also provides mechanisms to ensure the appropriate control and acknowledgement of messages.

Layer 4 — the transport layer. This is the terminal-to-terminal layer. Data may be carried across the networks using various forms of Layers 1, 2 and 3 (e.g. via a LAN and ISDN) but the terminals must have information at appropriate rates.

Layer 5 — the session layer. This defines the way in which applications running at the two ends of the link intercommunicate, including initiation and termination of sessions and co-ordinating their activity during the session.

Layer 6 — the presentation layer. This establishes the common format which is to be used between terminals, using common rules for representing data.

Layer 7 — the application layer. This is the task to be performed, for example file transfer, airline booking, message handling.

The lowest three layers are called the network service layers. In the ISDN there are two types of channel. First the 64 kbit/s 'B' circuit-switched channel for customer use. For straightforward data purposes in simple terms the ISDN only defines the Layer 1 attributes of this channel, the customer being free to use the bits provided for his own higher layer protocols. The 16 kbit/s 'D' signalling channel includes all three layers and these are discussed in Chapter 5.

The usual diagram for describing these sets of protocols is shown in Figure 1.7. The layers are stacked vertically. Two stacks are shown, the control 'plane' for signalling and the user 'plane'. The lines passing vertically through the centre of each stack show the progress of the relevant channels. The user information line carries applications (Layer 7) from terminal to terminal, only changing its physical form (Layer 1) at intermediate points such as the NT1 or local exchange. However, the control information in the form of signalling has only three defined layers. As will be seen in Chapter 5, Layers 2 and 3 of the protocol are terminated in the exchange. This is much simplified but illustrates the type of diagram

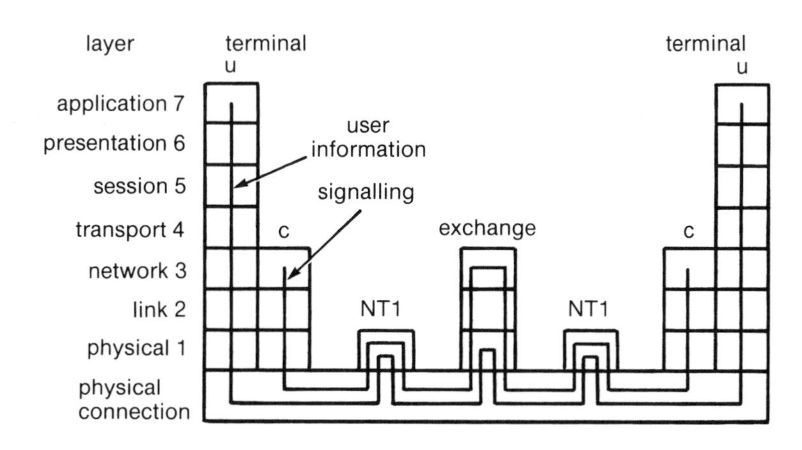

Figure 1.7
7-layer stacks.

used to illustrate message flows. A further complication arises from the fact that the 7-layer model is in no way absolute and sub-layers can often be identified. Hence diagrams with dotted intermediate levels are often encountered. Similarly other 'planes' are possible.

REFERENCES

The CCITT Recommendations are published by the International Telecommunication Union, Geneva.

The OSI 7-layer model is described in the International Standards Organization (ISO) Standard 7498.

QUESTIONS

1 What are the advantages of digital transmission over analogue transmission? In view of these advantages why did the analogue telephone network predominate for 100 years?

2 If the signal power equals the noise power in a channel of bandwidth 1 Hz, what is the theoretical information rate in bit/s which can be carried through this channel?

3 A telephone, data terminal (with modem) and facsimile machine can all be connected in parallel on an existing analogue public switched telephone network. What advantages would be obtained by connecting:
(a) the non-voice terminals to a data network?
(b) all the terminals to the ISDN?

4 Two users make arrangements to exchange information. Initially they use a data terminal via a modem on the public switched telephone network. They then repeat the same exercise via a packet network. Which of the ISO 7 layers are the same in both cases?

Chapter 2

Pulse Code Modulation

The International Standard ISDN is based upon 64 kbit/s circuit-switched channels and transmitted through a network which, for the foreseeable future, will predominantly be used for telephony purposes. Thus, although the great opportunities offered by ISDN may be in the fields of data and information technology, it is appropriate to consider the *raison d'être* of the network on which it is founded. The 64 kbit/s channel is regarded as virtually fundamental, but in some ways is a historical accident. As mentioned in the previous chapter, PCM was conceived by A. H. Reeves but, apart from some military applications in the Second World War, the first application was developed in the USA in the early 1960s for increasing the capacity of existing copper pairs between switching nodes. A diagram of the modulation process involved is shown in Figure 2.1.

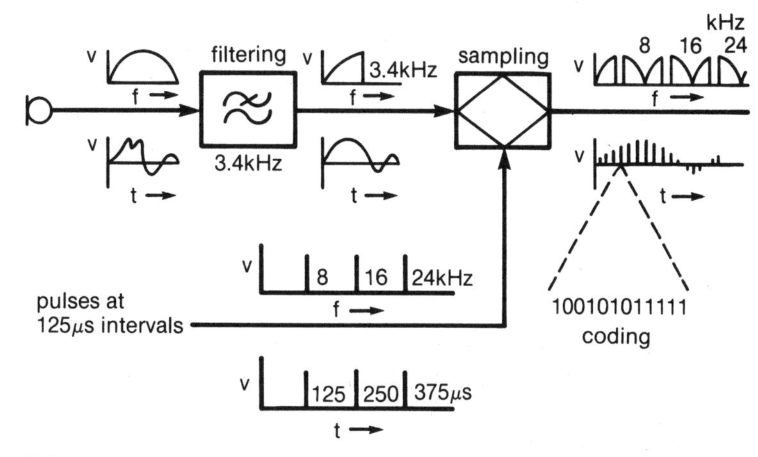

Figure 2.1
PCM coding.

2.1 SAMPLING

The analogue signal is band limited to 3.4 kHz and sampled 8000 times per second. This sampling rate arises from the process which gives rise to sidebands around frequencies which are integer multiples of 8 kHz as shown in the diagram. If the sampling rate is less than twice the bandwidth of the analogue signal, then the sidebands will overlap (known as 'aliasing') and subsequent decoding is not possible. This is Nyquist's sampling theorem. Note that there is another strong reason for sampling at a multiple of 4 kHz, owing to the fact that at the time of introduction of the PCM systems there existed Frequency Division Multiplex (FDM) systems in the network which carry telephony on sinusoidal carriers at multiples of 4 kHz. Inevitably there is some leakage of the PCM sampling waveforms into the FDM systems which could give rise to whistles unless they were multiples of 4 kHz, in which case the beats would appear as very low frequencies and be inaudible.

2.2 CODING

The amplitude of the samples is then measured and encoded into a binary number. Obviously the precision with which this encoding process can take place depends on the number of bits in the binary number. Any lack of precision is known as quantizing distortion. Empirically it was found that to give adequate speech quality 12 bits were needed in this encoding process. That is to say ± 2048 levels could be identified. There is an economy that can be made in these bits. With the straightforward 12-bit encoding the distortion would be the same whether the speaker was whispering or shouting. However, at higher speech levels much greater quantizing distortion can be tolerated as it is masked by the speech itself.

2.3 COMPRESSION

To exploit this masking a non-linear compression function is introduced to translate the 12-bit code into an 8-bit code. The ideal form of compression is logarithmic as this characteristic leads to a constant signal-to-quantization distortion ratio. This is shown in Figure 2.2 which shows diagrammatically how the uncompressed code could be mapped on to a compressed code for both polarities of signal. The problem arises for very low level signals as the logarithm of very small numbers becomes negative. Somehow the compression law has to be forced to pass through the origin. Two solutions were found. In North America the two logarithmic curves were displaced towards the central vertical axis giving a transfer function of the form $y \propto \log(1 + \mu x)$. This is known as the 'mu-law'. In Europe a line was drawn tangentially to the two curves and hence, by symmetry, it passes through the origin. Thus the curve is of the form $y \propto Ax$ over the central part of the

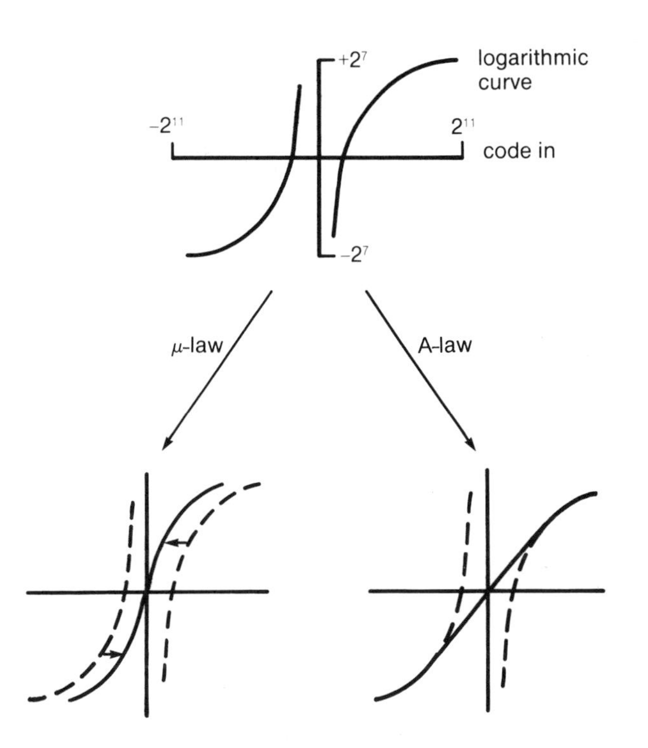

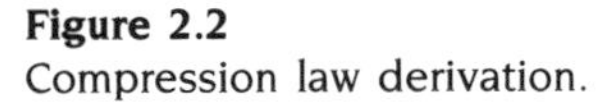

Figure 2.2
Compression law derivation.

range and $y \propto 1 + \log(Ax)$ at the extremes of the range. This is known as the 'A-law'.

To simplify further the translation process, the actual curves are approximated by linear segments each of half the slope of the previous. Although the theoretical derivation of the compression laws has been described, in practical terms the standard exists in the form of a tabulation for the conversion in CCITT Recommendation G.711. These are plotted in Figure 2.3. This reveals the exaggerations of Figure 2.2. In practice the two curves are very similar. Note that a symmetrical code is used so that the first bit indicates the polarity of the sample and subsequent bits indicate the magnitude of the sample. Additionally, in the European A-law case alternate bits in the code word are inverted; this is because in practice channels are frequently idle and, without the inversions, the idle signal would contain a lot of zeros which may make derivation of timing information difficult in line transmission systems. Similarly the North American system inverts all binary digits which increase the density of 1s for transmission purposes. The decoding process is the inverse of the encoding process and is again summarized in Figure 2.4.

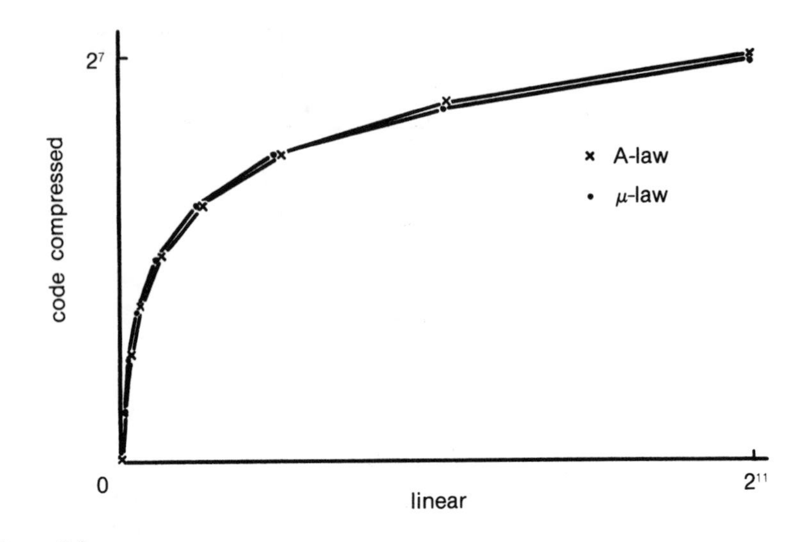

Figure 2.3
μ- and A-law companding.

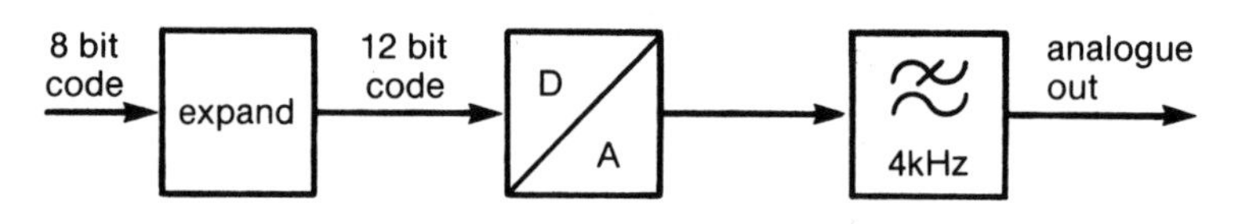

Figure 2.4
PCM decoding.

2.4 64 kbit/s CHANNELS

The process described above based on 8-bit coding of samples at 8 kHz gives rise to the 64 kbit/s fundamental channel rate. Earlier in the Chapter I referred to it as an historical accident. This is for two reasons:

(a) The A-law and the μ-law PCM Standards were chosen so that the transmission systems could be embedded anywhere in an analogue network. Owing to the need to ensure that analogue networks are stable, it is not possible to compensate for all the network attenuation and hence levels will vary by some 15 dB according to the location in the network. In an ISDN, speech encoding and decoding are always performed at the customer's terminal and the only level variation is due to speaker differences. Without the variations in level due to the network, only about 6 bits are required for PCM coding to give an adequate signal-to-quantization distortion. In addition the system is designed to permit several conversions from analogue to digital and back again with acceptable quantizing distortion.

(b) There are more efficient ways of encoding speech than PCM. With human speech, samples are not totally independent of one another. In fact there is considerable correlation between successive samples and encoders have been developed which exploit this fact. Previous samples can be used to predict the next sample, and then only the errors in the prediction have to be transmitted. Section 8.1 describes how this may be exploited.

For these reasons a digital telephony network could be constructed which offered telephony quality at 32 kbit/s or even lower. Nevertheless, the 64 kbit/s channel is likely to remain the fundamental for the terrestrial ISDN. The reasons are:

(a) Optical fibres offer large bandwidths.

(b) Switching and multiplexing costs tend to be related to the number of channels rather than the bit-rate.

(c) VLSI components have been developed for use at 64 kbit/s.

(d) 64 kbit/s gives a degree of headroom above modem data rates which should suffice for a reasonable period, and more sophisticated encoding techniques can be used to give higher quality speech.

(e) Avoiding interworking problems between parts of the network operating at different rates. The one standard channel size means that clear channel working between terminals can easily be guaranteed. Mobile systems operating within the bandwidth constraints of free space will, however, use the low bit-rate coding.

REFERENCE

CCITT Recommendation G.711. Pulse Code Modulation (PCM) of voice frequencies.

QUESTIONS

1 Voice frequencies are limited to 3.4 kHz. Give two reasons why a PCM sampling rate of 6.8 kHz is not used.

2 Music requires a bandwidth of 15 kHz. Experience has shown that the simple compression technique described for speech is not satisfactory. If it were decided to

use linear coding with an overall coding accuracy of 0.03% of the peak amplitude, what overall bit-rate is required?

3 Why is it not possible for an A-law or μ-law PCM encoder to continuously generate a long string of 0s on the transmission system?

Chapter 3

The Integrated Digital Network

The Integrated Digital Network consists of local switching nodes (usually called exchanges in Europe and central offices in North America) joined by digital transmission links carrying 64 kbit/s digital channels multiplexed on to a copper pair, coaxial pair, microwave or optical fibre bearer. The switching nodes are controlled by processors which intercommunicate through signalling links. In this chapter I will cover in outline these matters, but only in so far as they affect the ISDN.

3.1 MULTIPLEXING

The advantage of encoding speech into PCM is that it may be processed as would any other data stream, using digital integrated circuits. The multiplexing process involves taking the PCM blocks of 8 bits which are generated every 125 μs (i.e. 8000/sec) and interleaving them with blocks from other PCM encoders to give a Time Division Multiplex (TDM) of channels. Of course, in an IDN the 64 kbit/s sources need not be speech encoders but could be other data sources. In the 1970s North America and Europe chose different ways of assembling the multiplex. North America interleaved 24 channels giving an aggregate rate of 1.536 Mbit/s (24 $\times$ 64 kbit/s). At the start of each frame of 24 channels a marker bit F, is included which adds a further 8 kbit/s to give a total rate of 1.544 Mbit/s. This is shown diagrammatically in Figure 3.1. The F bit follows a defined sequence so that it is unlikely to be imitated by data in the traffic channels. The 24-channel PCM systems are widely used for connections between central offices in North America and for this purpose an 'associated channel' signalling system is used where in every 6th frame the least significant bit of the 8 bits in the channel is 'stolen' for signalling relating to that particular channel. Although a convenient system for the original point-to-point purpose this does mean that the channel available is not the full 64 kbit/s for ISDN purposes, and also is not readily adaptable to common channel interprocessor signalling. For these reasons in an ISDN

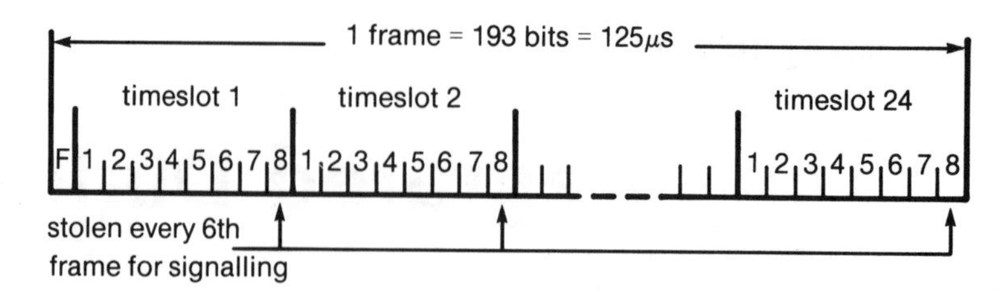

Figure 3.1
Frame structure of 1.544 Mbit/s interface.

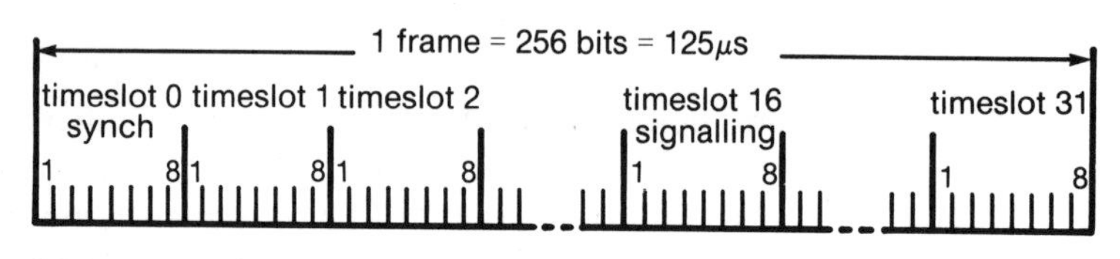

Figure 3.2
Frame structure of 2.048 Mbit/s interface.

environment the 'bit-stealing' process is being abandoned and the 24 channels are available for full 64 kbit/s transmission. Signalling is carried as one of the 64 kbit/s streams leaving 23 traffic channels.

In Europe a different format was adopted. Thirty 64 kbit/s channels were interleaved with a further 64 kbit/s signalling channel and a 64 kbit/s synchronizing channel. This gives an aggregate rate of 2.048 Mbit/s. This is shown diagrammatically in Figure 3.2. Channel 0 is used for the synchronizing (and also some maintenance) channel. Channel 16 is used for signalling. Once again in the original form the signalling channel had fixed bits allocated to particular traffic channels (i.e. channel associated), but in an ISDN environment channel 16 is used for a common channel signalling system.

The 1.544 Mbit/s and 2.048 Mbit/s are known as primary rate multiplexes. Further interleaving of several primary rate multiplexes can take place giving rates of 6, 45 and 274 Mbit/s in North America and 8, 34 139, and 560 Mbit/s in Europe. As far as the user of the ISDN is concerned, such rates are not visible and will not be discussed further until Chapter 9.

3.2 TRANSMISSION

It was mentioned in Section 1.6 that the earliest use of PCM was to increase the capacity of links on copper pairs between switching nodes. Previously two pairs of copper wires could carry one audio frequency telephone conversation if the two directions of transmission were separated, or two audio frequency telephone conversations if the two directions of transmission were on the same pair (but this presents problems with amplification as mentioned in Section 1.2). The two pairs could carry 24 (North America) or 30 (Europe) conversations if primary rate PCM systems were used.

However, transmission at 1.544 or 2.048 Mbit/s gives rise to two main problems:

(a) *Attenuation*. The loss of typical pairs is of the order of 15 dB/km at the relevant frequencies.

(b) *Crosstalk*. There is no screening between the pairs in the cable. They rely on their symmetry and various twists to prevent a signal in one pair from being coupled to another pair.

For these two reasons regeneration of the signal is required approximately every 2 km. In fact the argument was actually taken in reverse. The PCM systems replaced audio systems whose transmission performance was improved by periodically inductively loading the copper pair at intervals of 6000 feet. Hence additional underground capacity and cable access was available at these sites which could also be used for PCM regenerators. Working backwards through the cable characteristics, this defined the transmission rates chosen. Europe chose its higher rate several years after North America and indicates a degree of optimism based on experience of earlier systems.

The simplest way of transmitting binary digits is to represent the 1 and 0 by two voltage levels. For line transmission this simple process is rarely used for two reasons:

1. *Maintenance of balance*. It is convenient to include a high-pass filter at the input and output of regenerators so that power may be fed along the same pair and separated from the signal. This high-pass filter is usually a transformer. Transformers also give protection against surges. If long strings of signals of the same polarity are transmitted then after passage through the high-pass filter they will be severely distorted. The simplest way of balancing the signal is to transmit '0' as 0 V and '1' as alternately a positive and negative signal as shown in Figure 3.3. This is known as 'bipolar', 'pseudo-ternary' or 'alternate mark inversion (AMI)'.

2. *Timing extraction*. The regeneration process always involves retiming the signals and this is done by averaging the transitions over a considerable period and then using this extracted clock to retime subsequent pulses. In early regenerators the AMI signal was simply rectified and applied to a low-loss tuned circuit. More recently phase locked loops are used. For this process to work a certain minimum density of transitions is required. The AMI line code does not ensure this in the case of a string of 0s.

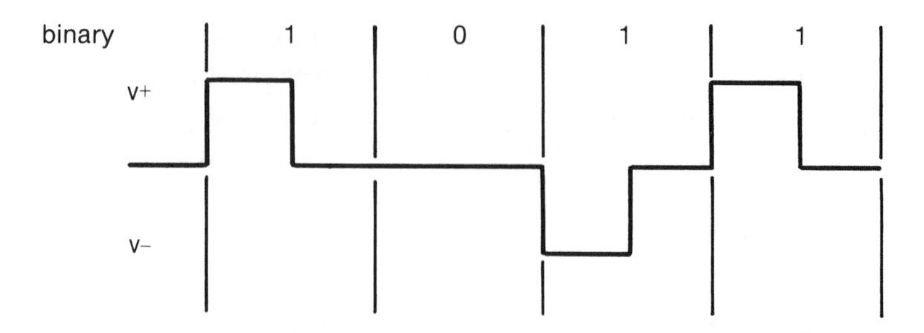

(note: the version of AMI used on the basic rate ISDN interface uses full width pulses and the binary is negated; see section 5.1.4)

Figure 3.3
AMI with half-width pulses.

For encoded speech this is rarely a problem taking the precautions mentioned in Section 2.3. However, the use of channels for data can give rise to such unfortunate patterns and the European systems and modern North American systems modify the basic AMI coding. In Europe the modified coding is called HDB3 (high density bipolar3). The basic AMI code requires that successive 1s are transmitted with alternating polarity. That is to say 10111 would be transmitted as +0−+−. If two 1s of the same polarity were transmitted then this would be a 'violation' of the rules. HDB3 operates by replacing any group of four 0s by a sequence consisting of three 0s followed by a violation of the coding rules (i.e. a symbol of the same polarity as the previous). Thus 11000001 would be sent as +−000−0+ or the inverse. The violation indicates the four 0s. It is possible with this simple rule that successive violations would be of the same polarity and a low frequency content added to the signal. To avoid this the four 0s can be substituted by B00V (B represents a normal bipolar pulse) to ensure that violations alternate in polarity; thus 100001100001 would be coded as +000+−+−00−+. Note that the 3 in HDB3 indicates the longest run of zero voltage symbols which can occur in the coded stream. In the North American system the modification to AMI is called B8ZS (bipolar with 8 zeros substitution). Whenever eight 0s occur they are coded 00B0VB0V. As this code contains two violations separated by an ordinary bipolar pulse it is inherently balanced. Up to seven consecutive 0s can be transmitted.

These coding techniques were developed in the 1960s when VLSI was not available and hence regenerators were simple by modern standards. With more complex technology the above constraints are much less important and in Chapter 4 it will be seen that coding effort is mainly directed to spectrum control, to minimize the effect of crosstalk and attenuation for local network applications.

The 1.544 Mbit/s and 2.048 Mbit/s are firmly established and in the future may be more relevant as interface Standards between equipment rather than as Standards for signals on copper pairs.

3.3 SWITCHING

If one had several terminals connected to individual pairs, which one wished to interconnect, one would arrange to connect the pairs to a switch matrix and close those cross points which allow the appropriate terminals to intercommunicate. Figure 3.4 shows a 6 × 6 matrix. The crosses indicate the closed switches and indicate that input terminal 1 is connected to output terminal 3, 2 to 4, 3 to 2, and so on. This is known as a space switch. The data from terminals could have been time division multiplexed on to a single pair, shown diagrammatically in Figure 3.5 by the rotating switch on the left. The data channels will now appear as sequential timeslots (each containing 8 bits) on the single pair. To connect to the appropriate terminals on the right-hand side needs a reordering of the channels and this is performed by the switch in the centre. The process involves putting the bits in each timeslot into memory and reading them out in the appropriate order. This function is called a time switch. The demultiplexing switch shown symbolically on the right will now be connected to the appropriate terminals on the left. In a real digital switch both space- and time-switching functions are required. Figure 3.6 shows a block diagram of a typical digital switch. So that the channels are in a convenient order 24 or 30 channel

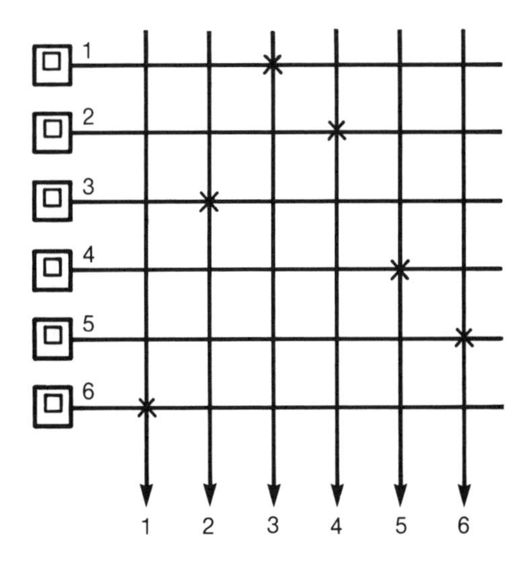

Figure 3.4
A space switch.

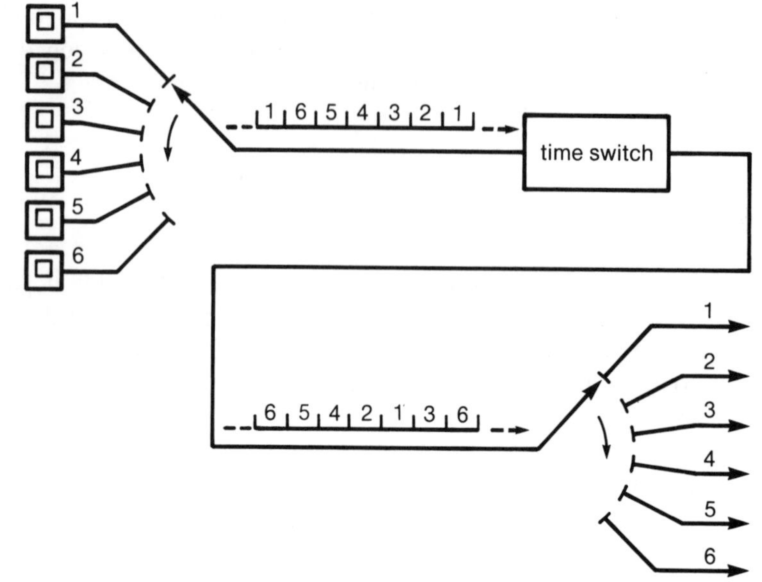

Figure 3.5
Time-switching of a time division multiplex.

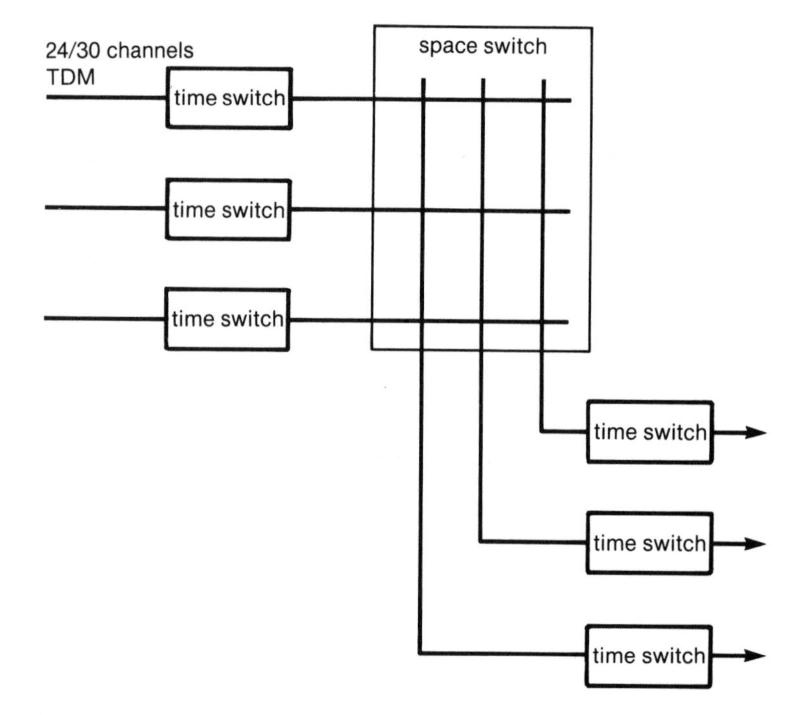

Figure 3.6
TST digital switch.

digital links are first time-switched individually. A space switch then allows timeslots to be transferred from digital link to digital link. Notice that during each timeslot there is a different configuration of the space switch so that each timeslot may be routed to a different digital link. After the space switching there is a further stage of time switching to reorder the channels as required. The most obvious economy in a time-divided digital switch compared with an analogue switch lies in the space switch, which is used in as many different configurations as there are timeslots. In the example of Figure 3.6 the space switch is being used to switch $3 \times 24 = 72$ or $3 \times 30 = 90$ channels instead of the three channels it could switch standing alone as an analogue switch. For this reason in real digital switches there is an additional stage of multiplexing of digital links to increase the number of channels time division multiplexed and offered to the space switch. The switch described has a Time–Space–Time structure (TST). Other structures are possible but the TST now predominates.

3.4 REMOTE UNITS

Figure 3.7 shows the basic components of an exchange. The concentrator is a switching stage which connects the subscribers to the number of internal trunks which are needed to carry the traffic that the customers generate. Typically the number of internal trunks may only be one tenth of the number of customers. The route switch connects the call to an external trunk or back to the concentrator for own exchange calls. The whole is controlled by the processor.

The use of digital transmission means that impairments which were introduced by analogue links no longer exist. Processor control and signalling links mean that the concept of an exchange as a single indivisible piece of equipment is not essential. Instead, if economics support it, pieces of the

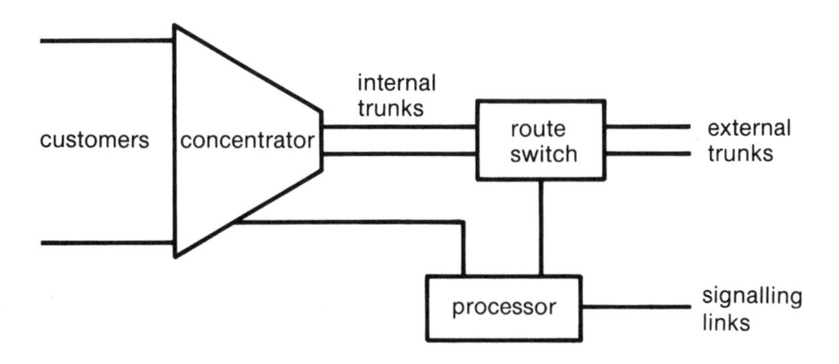

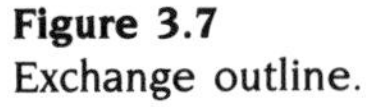

Figure 3.7
Exchange outline.

exchange may be separated. For example the processor may control several concentrator and route switches at remote sites. Similarly parts of the concentrator may be separated. One form which is widely used in North America consists of a unit which may be remotely sited from the exchange, possibly in 'street furniture' such as roadside cabinets, and typically has 96 line cards to which customers may be connected. The customers are then multiplexed on to four 1.544 Mbit/s bearers. These may then be regarded as a replacement for the 'front end' of the concentrator stages. Such systems when used for telephony purposes are often known as 'digital loop carrier' systems. They have the advantage of concentrating 96 customers on to eight copper pairs (four in each direction) or two optical fibres, which is particularly advantageous where cable pairs are scarce and difficult to provide. It is also a means of providing service to a small community, where reach considerations mean that they cannot be served directly. The digital links from remote unit to exchange may be of any length provided appropriate regeneration is provided.

A further use of remote units which is particularly relevant in an ISDN context is to introduce a service in advance of that service being generally available. Two aspects are significant.

(a) *Out of area service.* In the early days of the ISDN only some local exchanges may be suitable (e.g. the digital ones). Customers on other exchanges may be served by small remote units until their local exchange is replaced by a suitable exchange. This is how the British Pilot ISDN service (see Figure 1.5) was introduced so widely in 1986.

(b) *Limited demand.* In the early days of ISDN the demand will inevitably be small in comparison with the demand for conventional analogue telephony and it may not be economical to go to the complexity of modifications to the whole exchange for the sake of a small number of ISDN connections. In this case a remote unit may provide the ISDN interface to the small number of initial customers, putting off the more extensive exchange changes until demand is more significant. This has a further advantage that when tracking changes in International Standards only a small piece of equipment need be modified, rather than the whole exchange. In this case the 'remote' unit may actually be co-located with its parent exchange.

A further development of the remote is the 'adjunct' switch. This could be used at the front end of an analogue exchange. Any ISDN voice calls are routed to the analogue exchange, but calls requiring 64 kbit/s are routed to a digital network.

3.5 SIGNALLING

All modern switches are controlled by a central processor. These are usually called Stored Program Controlled (SPC) exchanges. Information flows between processors by means of Common Channel Signalling systems (CCS) and their use is fundamental to the provision of an IDN and hence an ISDN. This title distinguishes them from earlier systems where the signalling was carried along with the speech channel to which it related, which were called Associated Channel Signalling systems (ACS). The disadvantage of ACS in an SPC environment is that each channel into the exchange (of which there may be several thousand) has to have associated with it some form of processor access. CCS only needs, in principle, as many links as there are processors with which it needs to communicate. In practice these links will be replicated for traffic and reliability reasons.

Early CCS systems were CCITT No 6 in Europe and the similar North American CCIS (Common Channel Interoffice Signalling). For ISDN purposes these are now being superseded by the CCITT Signalling System No 7. This is primarily intended to be carried between processors on a 64 kbit/s bearer. It has a layered structure, similar to, but not the same as, the OSI layered structure described in Section 1.10. The lower levels are called the Message Transfer Part (MTP) whose purpose is to transfer signalling messages over the network in a reliable manner. It has a 3-level hierarchical structure.

Level 1 — encompass the physical signalling data link, which in the digital network consists of a 64 kbit/s timeslot in a digital multiplex. In principle any timeslot can be used but in the European system it is traditional to use timeslot 16; in the North American system timeslot 24 is used.

Level 2 — comprises functions which facilitate the reliable transfer of messages over a single signalling link. These functions ensure that messages are delivered in the correct order, without loss or duplication. To each message this level adds 16 check digits whereby the equipment receiving the message can check whether any errors have been introduced. Sequence numbers are also added to identify the current message and to acknowledge those messages which have already been received.

Level 3 — ensures the reliable transfer of signalling messages even in the event of network failures. Functions and procedures are included which inform the remote parts of the signalling network of the consequences of failures and reconfigure the routing of messages through the signalling network to overcome them.

Carried by the MTP, and hence hierarchically above the MTP, are the user parts. These are the messages which are used to set up, clear down and otherwise control calls. Included in this area is the facility to access non-switching nodes such as intelligent network databases, and also to support network operations and maintenance activities.

The user part for controlling setting up telephony calls is called the Telephony User Part (TUP). In Britain there is a national variant of this known as National User Part (NUP). For ISDN purposes a new user part has been defined called Integrated Services User Part (ISUP) but as an interim measure in Europe an enhanced version of the TUP, called TUP+, will be used for early ISDN purposes. Ultimately it is planned that ISUP will supersede the earlier systems. Examples of some ISUP messages are:

Initial Address Message (IAM). A message sent in the forward direction to initiate seizure of an outgoing circuit and to transmit number and other information relating to the routing and handling of a call.

Subsequent Address Message (SAM). A message that may be sent in the forward direction after an initial address message, to convey additional called party number information.

Information Request Message (INR). A message sent by an exchange to request information in association with a call.

Information Message (INF). A message sent to convey information in association with a call requested in an INR message.

Address Complete Message (ACM). A message sent in the backward direction indicating that all the address signals required for routing the call to the called party have been received.

Call Progress Message (CPG). A message sent in the backward direction indicating that an event has occurred during call set-up which should be relayed to the calling party.

Answer Message (ANM). A message sent in the backward direction indicating that the call has been answered. This message is used in conjunction with charging information in order to: (1) start metering the charge to the calling customer, and (2) start measurement of call duration for international accounting purposes.

Facility Request Message (FAR). A message sent from an exchange to another exchange to request activation of a facility.

Facility Accepted Message (FAA). A message sent in response to a facility request message indicating that the requested facility has been invoked.

Facility Reject Message (FRJ). A message sent in response to a facility request message to indicate that the facility request has been rejected.

User-to-User Information Message (USR). A message to be used for the transport of user-to-user signalling independent of call control messages.

Call Modification Request Message (CMR). A message sent in either direction indicating a calling or called party request to modify the characteristics of an established call (for example, change from data to voice).

Call Modification Completed Message (CMC). A message sent in response to a call modification request message indicating that the requested call modification (for example, from voice to data) has been completed.

Call Modification Reject Message (CMRJ). A message sent in response to a call modification request message indicating that the request has been rejected.

Release Message (REL). A message sent in either direction, to indicate that the circuit is being released because of the reason (cause) supplied and is ready to be put into the IDLE state on receipt of the release complete message. In case the call was forwarded or is to be re-routed, the appropriate indicator is carried in the message together with the redirection address and the redirecting address.

Release Complete Message (RLC). A message sent in either direction in response to the receipt of a release message, or if appropriate, to a reset circuit message, when the circuit concerned has been brought into the idle condition.

Signalling will be discussed further in Chapter 5.

REFERENCES

CCITT Recommendations:
G.702. Digital hierarchy bit-rates.
G.703. Physical and electrical characteristics of hierarchical digital interfaces.
Q.700–Q.795. Specification of Signalling System No 7.

American National Standards Institute:
T1.102. Digital hierarchy—electrical interfaces.
T1.111–116. Signalling System No 7.
T1.609. ISDN Digital Subscriber System No 1 to Signalling System No 7 interworking.

Bell Communications Research: TR-TSY-000057. Functional criteria for digital loop carrier systems.

QUESTIONS

1 What percentage of transmitted bits are allocated to signalling on 24 and 30 channel PCM systems?

2 A sequence of binary digits 1000010000011000000001 ... is encoded in HBD3 as +000+− ... Complete the coding. How would the binary sequence be encoded in B8ZS?

3 Time switches are much less costly to build than space switches because of the availability of VLSI. Why then are space switches incorporated into switching nodes?

4 The local exchange consists of three main parts - a concentrator, route switch and processor. What are the merits of locating them all in the same place, and what are the merits of separating them?

Chapter 4

Local Network Digital Transmission

Peter Adams

In Section 1.7 it was highlighted that a fundamental problem of establishing an ISDN was in making use of the existing local network copper pair infrastructure to carry channels at 64 kbit/s. In this chapter we shall consider the problems and how they can be overcome.

4.1 THE LOCAL NETWORK

The local network has a tree structure with cables radiating from the local exchange. The cables branch at permanently connected joints and at cross-connection flexibility points. A typical section of a European local line network is shown in Figure 4.1. It is often referred to in two parts—main and distribution—the dividing line being the primary cross-connection point. The main network consists of pressurized cables and the distribution network of petroleum jelly filled cables, each thereby being resistant to the ingress of moisture. A variety of cable sizes and conductor gauges exist; generally the larger-size, thinner-gauge cables being used in the main network. Figure 4.2 shows an alternative way of providing flexibility by arranging that pairs may appear at several distribution points. This is more common in the USA and has dramatic effects on attenuation and impedance. The planning limits in force when the network was laid out varied. The attenuation limit is variously specified by different administrations as being 6 dB at 800 Hz or 10 dB at 1600 Hz or corresponding figures for intermediate frequencies. More modern exchanges may have limits of 10 dB at 800 Hz and 15 dB at 1600 Hz. The resistance limit depends on the signalling limits for the exchange. This ranges from about 800 Ω for early exchanges to 2000 Ω for modern exchanges. The cumulative distribution of the UK and USA networks is shown in Figure 4.3. Loop resistance is not of primary interest in determining the transmission characteristics of a cable although it does correlate well with cable loss especially for 0.4 mm (26

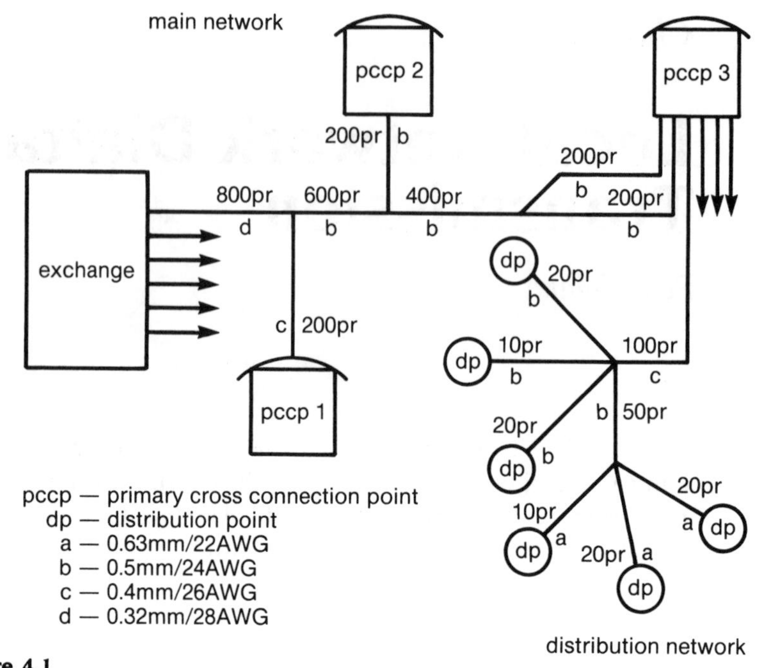

Figure 4.1
Local line network with flexibility points.

AWG) and 0.5 mm (24 AWG) pairs. However, subscriber loops are often required to be line powered and under maximum voltage limitations the loop resistance will affect the amount of power available at the subscriber's end. The changes in planning limits will cause a very gradual upward drift in average loop length which has to be borne in mind in the design of future transmission systems. Thus, in terms of structure and types of cable, the local network is extremely variable and so the electrical characteristics which affect digital transmission also vary dramatically from connection to connection.

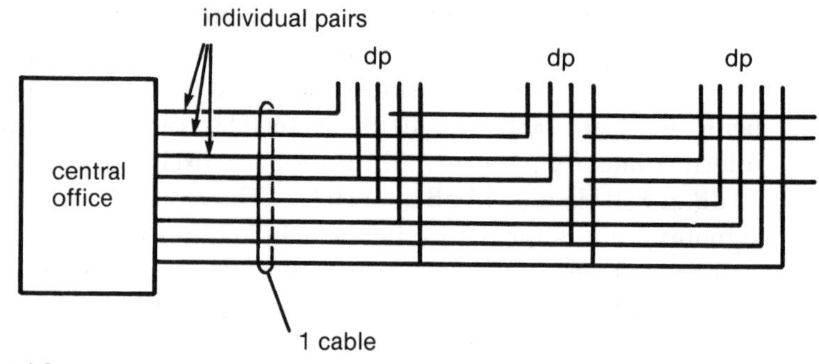

Figure 4.2
Local line network with multiple teeing.

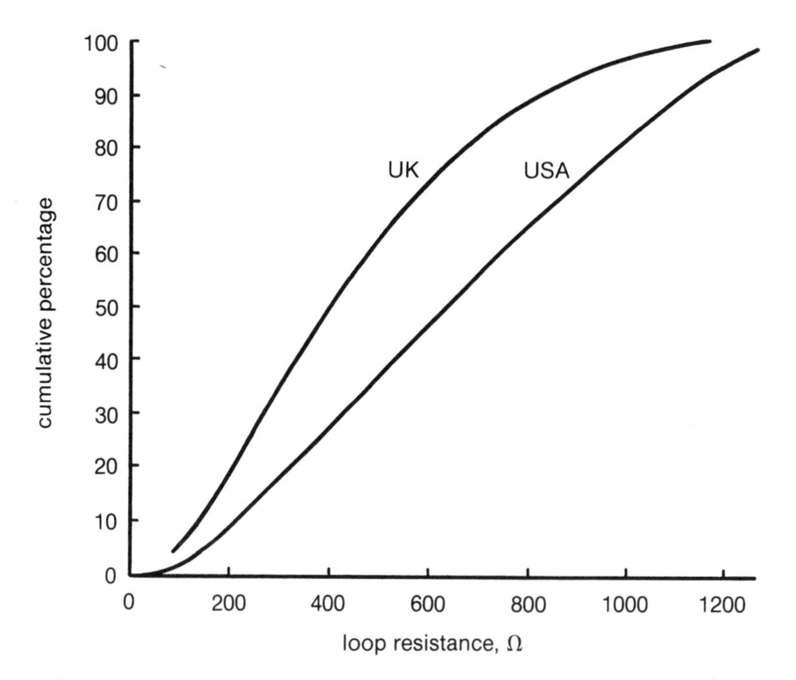

Figure 4.3
Cumulative distribution of loop resistances.

4.2 CABLE CHARACTERISTICS

The characteristics of particular interest for the design of digital transmission systems are:

(a) *Attenuation*. The variation of attenuation with frequency is plotted in Figure 4.4 for the gauges of conductor found in the local network. At frequencies below about 10 kHz and above about 100 kHz the attenuation varies with the square root of the frequency; in between there is a reduced attenuation slope. The attenuation decreases with increasing conductor gauge.

(b) *Phase*. The phase frequency characteristics of the various gauges of cable are shown in Figure 4.5. All the cables exhibit a linear phase characteristic with a constant phase offset above 10 kHz but are non-linear at lower frequencies.

(c) *Characteristic impedance*. The modulus of the characteristic impedance of the various cable types is shown in Figure 4.6. At higher frequencies it approaches a constant resistive value but at lower frequencies varies markedly with frequency and from cable to cable. The phase of the

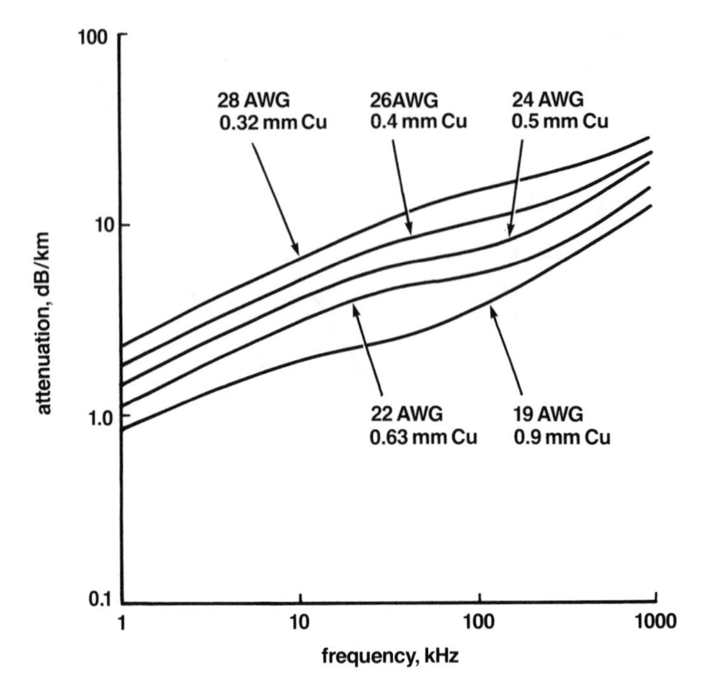

Figure 4.4
Attenuation of typical cables of various gauges.

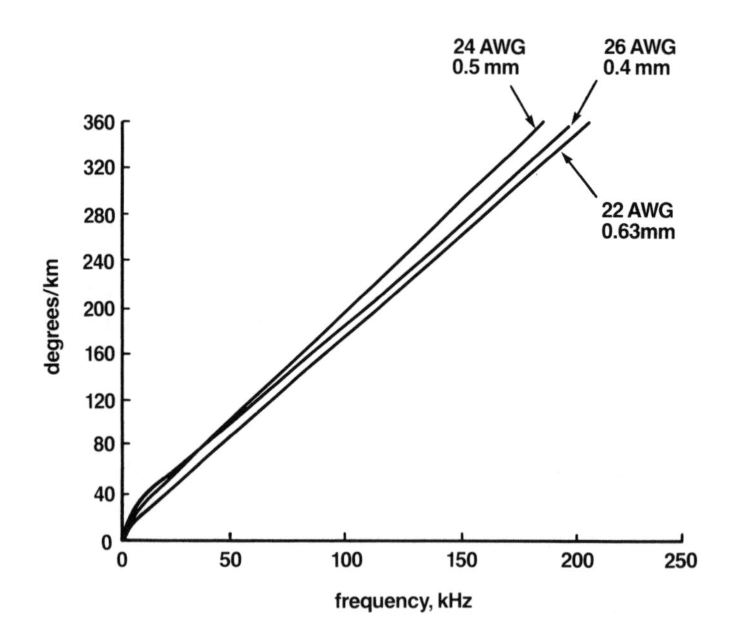

Figure 4.5
Phase coefficient of copper pairs.

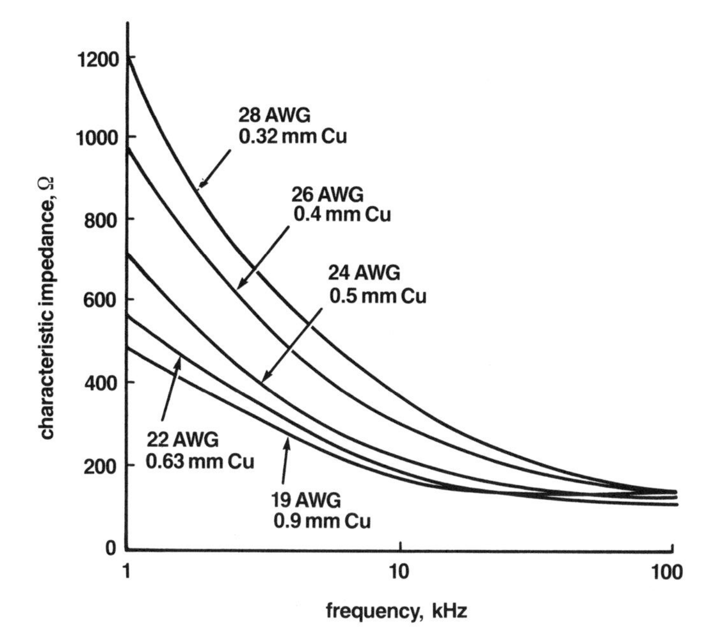

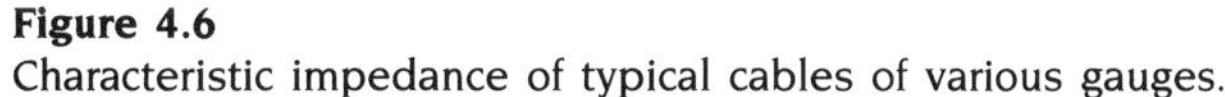

Figure 4.6
Characteristic impedance of typical cables of various gauges.

characteristic impedance also varies with frequency at low frequencies but at high frequencies approaches zero.

(d) *Crosstalk.* Two cases of crosstalk are considered. First where the transmitter and receiver are at the same end of the cable, called Near End crosstalk (NEXT). Secondly where transmitter and receiver are at opposite ends of the cable, called Far End crosstalk (FEXT). The crosstalk characteristics vary markedly between pairs in the same cable and with frequency. For the situation where many systems operate over one multi-pair cable we can average the crosstalk characteristics and then it is found that the Near End crosstalk power sum attenuation follows the familiar $f^{1.5}$ law. For the cables of the UK network it has been found that 98% of all connections have NEXT couplings with a crosstalk power sum attenuation of better than 55 dB at 100 kHz. For FEXT the situation is complicated by the tree structure of the local network but fortunately FEXT is less of a problem than NEXT and can be ignored where NEXT is the limiting factor.

(e) *Impulsive and random noise.* Impulsive noise, arising from electromechanical switching transients in adjacent cable pairs and electromagnetic interference from nearby electric traction systems and power lines, is the most serious kind of noise found in the local network. It is very ill-defined and is usually characterized by peak amplitude and

Figure 4.7
Margin against decision errors (threshold) in mV required for 99% error-free seconds expressed as a cumulative distribution.

rate of occurrence. A few measurements have been made in the UK; the plot in Figure 4.7 shows the results obtained at one local exchange. The plot shows the cumulative percentage of measurements for which a given threshold was not exceeded for 99% for seconds during an approximately 10-minute measurement period. Random interference in the form of thermal noise is not significant in the local network. In certain geographical locations, however, radio interference could be a problem.

4.3 MODULATION

Digital transmission requires the modulation of the digital information on to a suitable electrical signal; the most common way of doing this is to use shaped pulses and to modulate them by the data. Thus a pulse would be sent normally to represent a '0' and would be sent inverted to represent '1'. More complex arrangements are possible and a particular solution is to take two binary digits so that the one binary digit would cause the shaped pulse to be sent erect or inverted: the other binary digit would cause the pulse to be sent normal amplitude or a larger amplitude. Thus two binary digits are encoded as one symbol which can have four states. The process is called Pulse Amplitude Modulation (PAM) and as four states are allowed

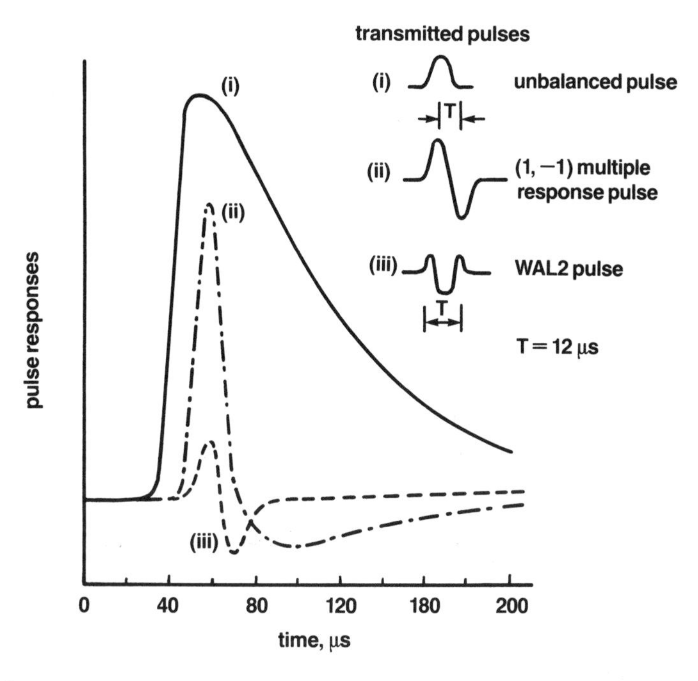

Figure 4.8
1200 loop pulse responses.

it is called quaternary. Modulated pulses are sent over the cable and, ideally, are detected without error at the far end. The cable characteristics result in pulse distortion which spreads the pulses ('symbols') and causes Intersymbol Interference (ISI), and crosstalk and noise which cause errors if the signal level is insufficient. The degree to which these factors affect transmission is a function of the line code and/or the pulse shaping used. Pulse shapes which have a spectrum concentrated at low frequencies suffer severe distortion because of the rapidly changing frequency response characteristics; however, they are relatively insensitive to crosstalk. Conversely, pulse shapes with little low frequency energy suffer little distortion but are more susceptible to crosstalk interference. This trade-off is clearly illustrated by Figure 4.8 where the responses of a typical cable to three pulse shapes are shown. The unbalanced pulse of width $2T$ seconds is severely dispersed but, as the spectral plot in Figure 4.9 shows, the spectrum peaks at zero frequency. The (1, −1) pulse suffers far less dispersion but is attenuated and has more spectral energy at higher frequencies. The WAL 2 pulse is virtually dispersion-free but suffers the highest attenuation and has a much wider spectrum.

The well-behaved frequency response at higher frequencies suggests that on shorter connections where crosstalk is not a problem (because of

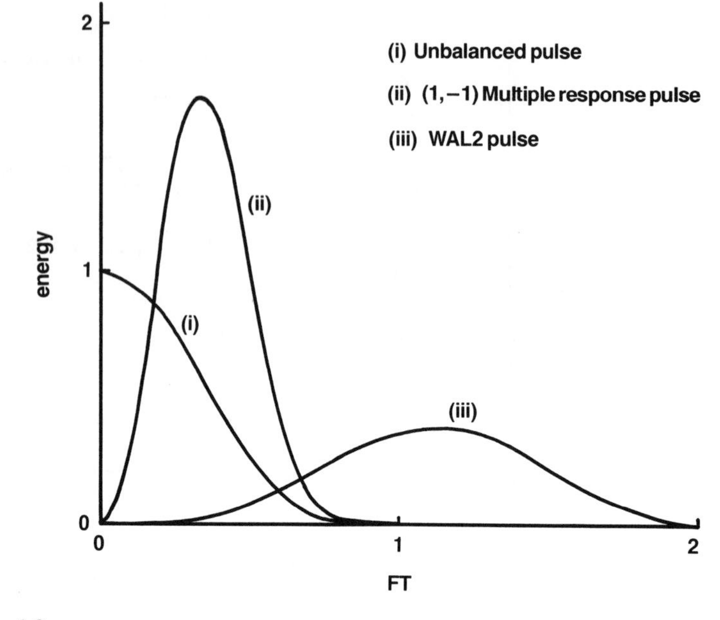

Figure 4.9
Energy spectra of pulses.

the lower line) modulation of the digital information on to a sinusoidal carrier can confer some advantages. At frequencies between about 10 kHz and 100 kHz the flatter attenuation response will introduce less distortion especially to a double-sideband amplitude modulated signal. Also, the process of demodulation will automatically remove the phase offset and the remaining linear phase introduces no distortion.

4.4 EQUALIZATION

As we have seen, the judicious use of pulse shaping or linear line coding can give varying degrees of reduction in dispersion and the resulting ISI. However, the more crosstalk-tolerant pulse shapes do suffer ISI and, because of the variability of the cable characteristics, the distortion is not easily predicted with precision. It is, therefore, necessary to use some form of adjustable (and preferably automatic) equalization of the cable frequency response. There are two fundamentally different ways of achieving equalization, illustrated in Figure 4.10.

(a) *Linear equalizer*. Linear equalization can be provided in two ways: by equalizing over the whole bandwidth of the line signal or, after

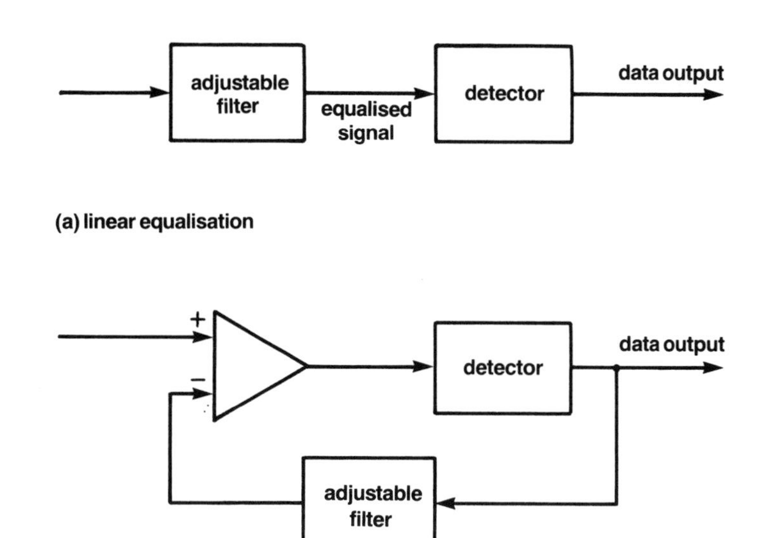

Figure 4.10
Types of equalizer.

sampling, over the sampled-signal spectrum. The former can be achieved by an adjustable continuous time analogue filter or by an adaptive Fractional-Tap (FT) discrete time equalizer in which the waveform is sampled at least at twice the highest frequency of the line signal; the latter by a T-spaced adaptive equalizer where the waveform is sampled at the symbol rate. By using an appropriate adaptation algorithm the FT equalizer is also capable of optimally filtering any noise present with the received signal. However, the T-spaced qualizer is easier to implement and, provided a good sampling phase is used, gives almost as good performance. Both forms of linear equalization introduce gain to combat the frequency dependent loss of the cable and so cause noise enhancement. A further disadvantage is that the processing required in the equalizer involves the multiplication of accurately represented quantities.

(b) *Decision feedback equalizer (DFE).* The DFE has the advantage of not enhancing noise because noise does not enter the filter. Also, because the filter is driven by detected data, if a transversal filter is used the multiplications involved are each between a coarsely quantized value and only one accurately represented value. However, because detected data is fed back, if an error does occur the resulting magnification, rather than cancellation, of ISI can cause error propagation.

The received pulse shapes of Figure 4.8 suggest that, as all the ISI can be made to arise from previously transmitted pulses, a pure DFE can be used to cancel it. However, the large, slowly decaying, response of the unbalanced pulse would mean that a long transveral filter would be required and the error propagation mechanism of the DFE would generate long bursts of errors. The unbalanced pulses, however, require a relatively short filter and the error propagation is far less severe. Effectively, the DFE can be used to advantage provided some compromise linear equalization of the cable is included. Line coding, or pulse shaping, is one way of doing this; alternatively a fixed linear equalizer at the receiver can be used.

4.5 TIMING RECOVERY

The form of timing recovery depends on the type of equalization adopted. If an ideal analogue equalizer is used then the receiver samples optimally at the peak of the resulting impulse response. For the FT linear equalizer any stable sampling phase will suffice as the performance is nearly independent of sampling phase. For the T-spaced linear equalizer a near-optimum sampling phase is given by a timing recovery scheme which achieved Band-Edge Component Maximization (BECM). The pure DFE requires that the pulse response is sampled so that any pre-cursor ISI is negligible.

4.6 DUPLEX TRANSMISSION

Digital subscriber loop transmission systems are required to operate in duplex mode over a single pair of wires. Theoretically, this is possible, as shown in Figure 4.11, by the use of a perfect hybrid balancing network. If Z_b is identical to the input impedance of a cable then the transmitted signal is completely removed from the input to the receiver. However, as we have already seen, the characteristic impedance of a cable is a complicated function of frequency and is difficult to match exactly. Also the coupling to line may be by a transformer creating a further impedance variation at low frequencies. Typically, a balance return loss of about 10 dB for low frequency pulse-shaped signals is all that can be achieved reliably with a simple compromise balance impedance. For modulated signals operating at a frequency where the characteristic impedance is nearly resistive typically a 20 dB balance return loss may be reliably obtained. Over all but the shortest cables the locally transmitted signal will seriously interfere with the received signal and so methods are required to remove the local echo (and any others from impedance mismatches on a mixed cable gauge connection)

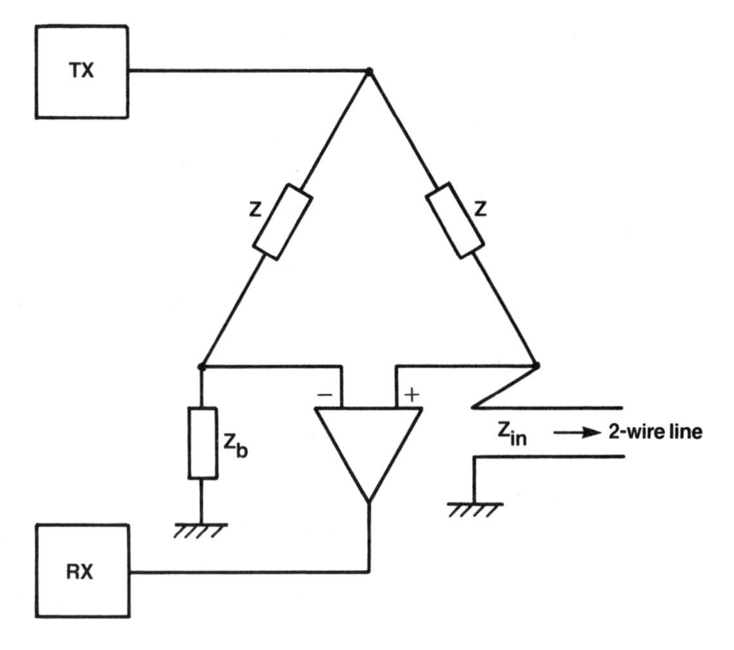

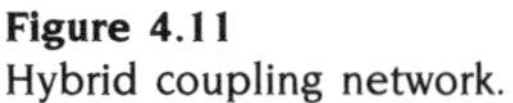
Figure 4.11
Hybrid coupling network.

from the received signal. There are three techniques available (Figure 4.12): frequency division, time division and echo cancellation.

(a) *Frequency Division Duplex (FDD).* In FDD the signals for the two directions of transmission occupy separate frequency bands and are separated by appropriate filters. The technique is very simple and inherently provides good tolerance to NEXT. However, the tolerance to FEXT and noise in general is degraded because the upper frequency channel suffers more attenuation. In addition, very high performance band-separation filters are required. These disadvantages have led to the exclusion of FDD as a candidate for digital subscriber loops.

(b) *Time Division Duplex (TDD).* Also known as 'burst mode' and 'ping-pong' transmission, this technique separates the two directions by transmitting and receiving in different timeslots as illustrated in Figure 4.12. The transmitted signal can be removed from the receive signal path by gating it out with an appropriately chosen timing waveform. As the received signal is not continuous the receiver timing recovery circuit has to either reacquire timing rapidly on each new burst or, exploiting the periodicity of the bursts, be heavily damped so that a 'flywheel' operation is possible. TDD is very simple to realize; no complex filtering is required and by using a well-balanced pulse shape no adaptive equalization is necessary. However, in order to accommodate duplex transmission, the bit-rate during a burst is at least twice the required continuous rate. Therefore, the tolerance to crosstalk and noise is degraded. It is argued that by synchronizing the bursts at the local exchange end, the effects of NEXT can be avoided but it is doubtful whether such a constraint is acceptable from an operational point of view and the degradation in tolerance to noise and FEXT still remains. Usually the burst bit-rate is significantly greater than twice the continuous bit-rate because of the effect of the propagation delay of the cable. If the propagation delay is t_g, the continuous bit-rate B, the burst bit-rate B_b and there are M bits per burst then B_b must be greater than $2MB/(M - 2Bt_g)$. If there are additional bits included in the burst for synchronization purposes then the value of M in the numerator is increased making the required bit-rate higher.

On shorter connections the burst mode technique offers a simple and effective means of duplex transmission. For this reason it is favoured for internal loops which terminate on PABXs. For public network use, however, the poor reliable transmission range makes it less attractive.

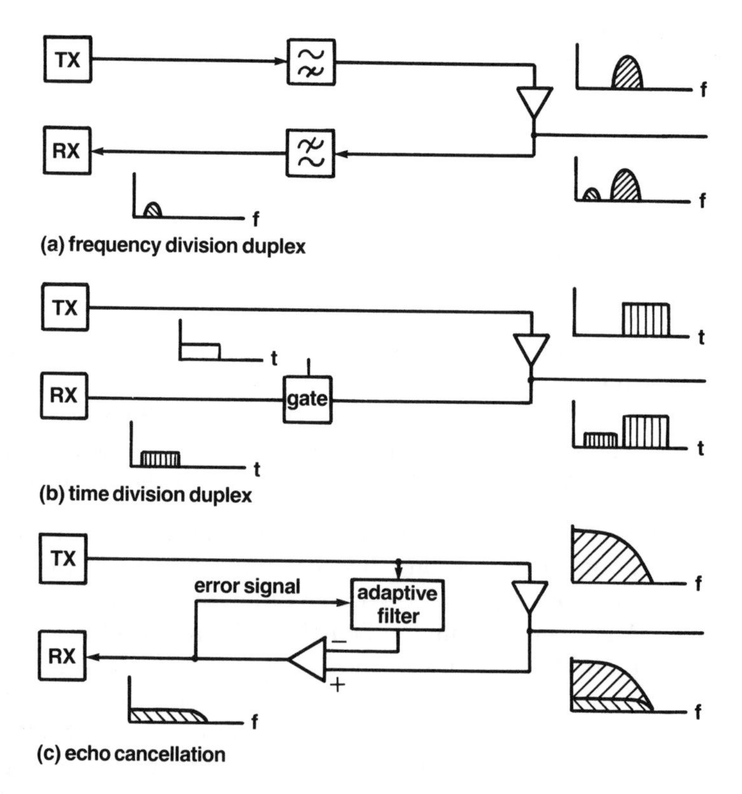

Figure 4.12
Duplex transmission techniques.

(c) *Echo cancellation.* Also known as the adaptive hybrid technique, echo cancellation involves adaptively forming a replica of the echo signal arriving at a receiver from its local transmitter and subtracting it from the signal at the input to the receiver. As shown in Figure 4.12 the signals sent in each direction can occupy the same frequency band and are continuous; therefore the disadvantages of the FDD and TDD techniques are avoided. The penalty is that an adaptive echo canceller requires a complex hardware implementation. However, modern VLSI technology enables this complexity to be implemented at acceptable cost so that it is now recognized that echo cancelling duplex systems offer the best solution to subscriber loops with longer reach.

4.7 A COMPARISON OF TRANSMISSION TECHNIQUES

Before comparing transmission techniques it is instructive to evaluate the limits to digital transmission in the local network. The fundamental limitation to the communications capacity is imposed by crosstalk. Applying Shannon's theory (see Section 1.4) the capacity as a function of cable length can be calculated. Using a value of NEXT power sum attenuation of 55 dB

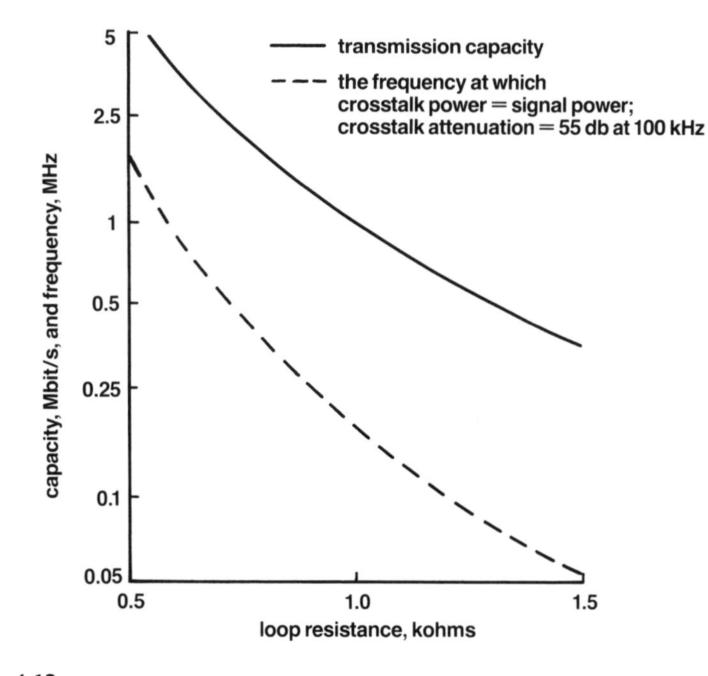

Figure 4.13

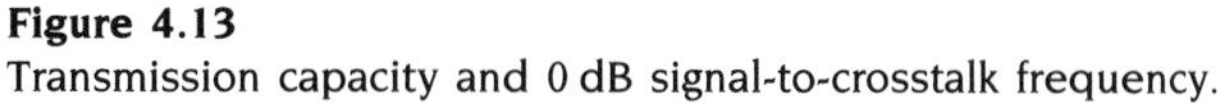
Transmission capacity and 0 dB signal-to-crosstalk frequency.

at 100 kHz and typical measurements of cable frequency response the graph in Figure 4.13 was plotted. The communications capacity of local network cables is clearly cable length dependent, but even over long lengths is sufficient to provide the data rates required for digital subscriber loops. Also plotted is the frequency at which the NEXT attenuation and cable attenuation are equal; this gives a bound on the maximum useful frequency above which more crosstalk is introduced than signal, suggesting that to achieve the highest performance the signal spectrum should be confined to low frequencies.

4.8 CROSSTALK ANALYSIS

Much effort has gone into the optimization of transmission systems in the local network. Making plausible assumptions about the local network it is possible to calculate the performance of equipment operating with different modulation schemes. Results are shown in Figure 4.14. The plots clearly show that pulse amplitude modulation is better than sine wave amplitude modulation, decision feedback equalization is better than linear

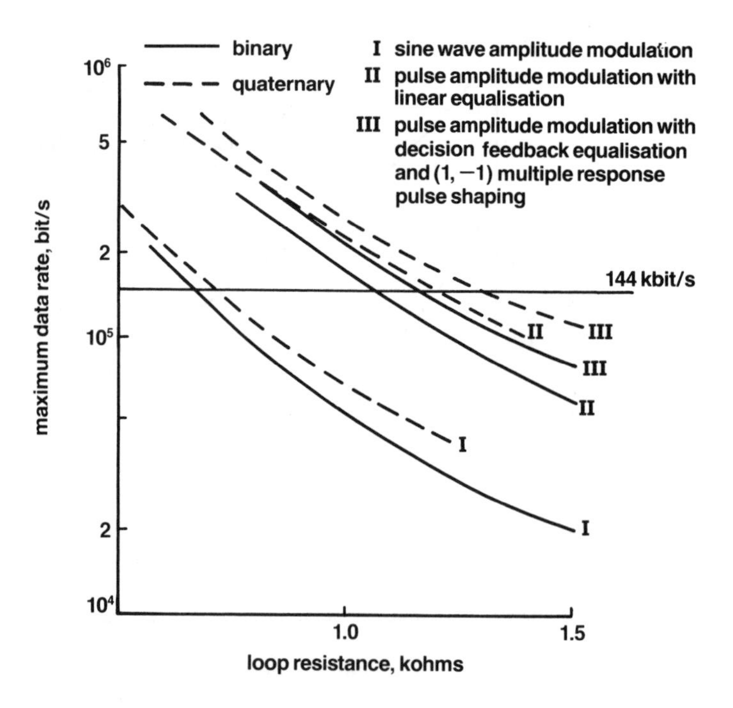

Figure 4.14
Crosstalk limited performance.

equalization and that multilevel coding allows greater data rates and/or greater reach.

4.9 IMPULSIVE NOISE

It is important that a system is designed to have the best tolerance to impulse noise. Such measurements of impulsive noise as have been done are usually in terms of the peak magnitude and frequency of occurrence, the implication being that impulses above a certain magnitude will cause errors. Unfortunately this simplistic approach neglects the effect of the receive filter and linear equalizer; to calculate the peak value of an impulse at their output a knowledge of impulse shape is required. In the absence of such information the best we can do is to compare systems by their tolerance to various carefully chosen impulse shapes.

Figure 4.15 shows the calculated tolerance of the various systems to a matched impulse when the transmit power is 10 mW into a 140 Ω load. PAM and decision feedback equalized systems have a clear advantage, but multilevel coding appears to decrease the tolerance to impulses. It should

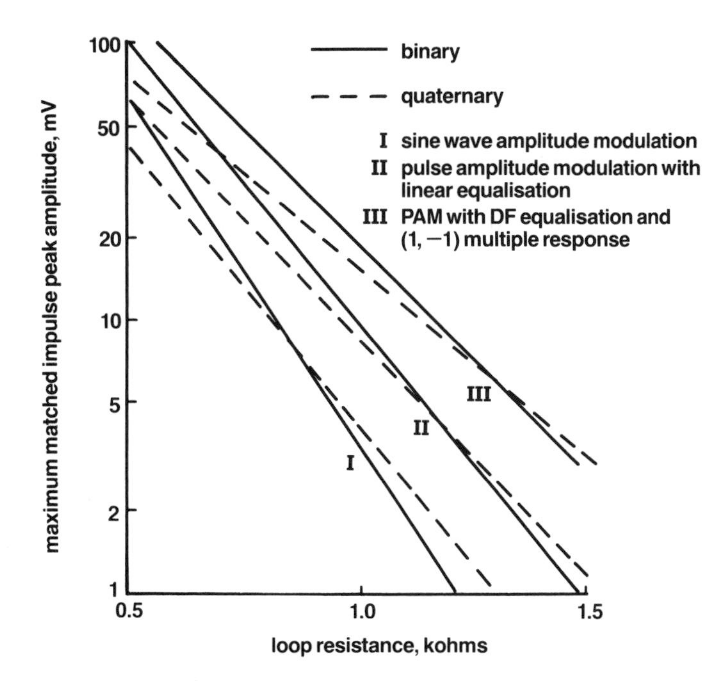

Figure 4.15
Maximum error-free impulse amplitude for 144 kbit/s transmission.

be noted, however, that the impulses required to cause errors in the four-level case have twice the energy of the impulses required in the two-level case. If we normalize the plots on an equal energy basis then the four-level systems are better than the two-level systems. This discussion highlights the difficulty of comparing systems if no precise knowledge is available on the shape of practical impulses.

4.10 ECHO CANCELLER STRUCTURES

Figure 4.12(c) shows a general echo canceller which operates on the output of the transmitter to produce an echo replica. The echo to be cancelled is a function of the current transmitted data, past data and the electrical characteristics of the transmitter, cable and the line interface. The latter normally result in an echo impulse response which decays with time and so the adaptive network can be constrained to model a finite duration channel only apart from some insignificant residual echo. Any filters in the transmitter will cause the amplitude of the input to the network to be of a continuous nature; simpler echo cancellers result if a data-driven structure is used as shown in Figure 4.16. The echo replica at time instant iT is given by

$$e(iT) = \mathrm{f}(d_i, d_{i-1}, d_{i-2}, \ldots, d_{i-N+1})$$

where d_i is the current data element and d_{i-j}, $j = 1$ to $N + 1$, are the past $N - 1$ data elements. In a binary system the d_i are binary valued; in a multilevel system of L levels the d_i are represented by B bits, where B is equal to or greater than $\log_2 L$. A straightforward implementation is to use the data elements as the address lines to look-up table which contains the 2^{NB} distinct values of $e(iT)$. An echo canceller realized in this way is known as a table look-up echo canceller and has the great advantage of simplicity. Also, there are no constraints on the function f(.). However, the implementation has two disadvantages: the memory size increases exponentially with N and, as we shall see later, it is slow to converge when made adaptive. Both these disadvantages are gradually diminished if the look-up table is subdivided to produce partial outputs $e_j(iT)$ which are summed to give $e(iT)$. Extending this process to look-up table per data element and then assuming a linear relationship between each value of d_i and its corresponding partial output e_j(IT), then a linear transversal filter echo canceller results; i.e.

$$e(iT) = \sum_{j=0}^{N-1} ej(iT) = \sum_{j=0}^{N-1} d_{i-j}.h_j$$

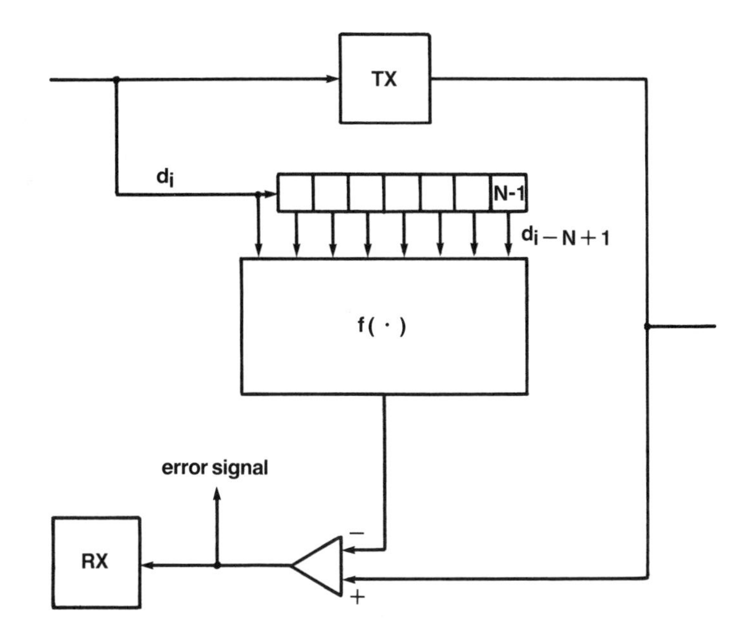

Figure 4.16
General data-driven echo canceller.

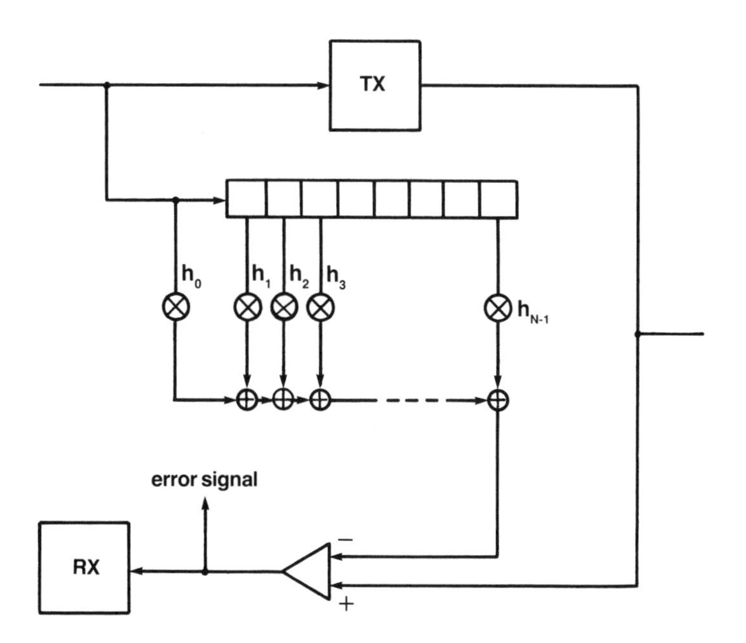

Figure 4.17
Transversal filter echo canceller.

where the values h_j are the samples of the echo impulse response. This structure is shown in Figure 4.17. The amount of memory required has been reduced to the minimum of N locations, but the processing increased to one multiplication and addition per delay line tap.

In the section on timing recovery it was noted that conventional timing recovery schemes require operations on the whole of the received signal. For such schemes to work correctly the echo must be removed from the whole of the received signal spectrum. Therefore, the echo canceller must operate at a sample rate that is at least twice the highest signal frequency in the received signal. Such fractional-tap spaced echo cancellers are easily obtained by multiplexing m (m greater than or equal to 2) of the structures of Figures 4.16 and 4.17 such that outputs are obtained every T/m seconds.

4.11 ECHO CANCELLER ADAPTATION

To adjust the echo canceller coefficients to their optimum values the error signal at the output of the subtractor is minimized by an appropriate adaptive algorithm. The most commonly used are the least-mean-squared (LMS) algorithms which seek to minimize the mean-squared error signal. If $e^*(iT)$ is the actual echo at time iT then the error signal is

$$r(iT) = e^*(iT) - e(iT).$$

In the look-up table case the algorithm is trivially simple; each location in the table is updated, when it is addressed by the algorith:

$$e((i + 1)T) = e(iT) + \mu.r(iT)$$

$e((i + 1)T)$ then replaces $e(iT)$ in the appropriate look-up table location. The gain constant (μ) determines the rate of convergence. For random data each location is addressed on average only 2^{-NB} of the time so convergence can be very slow.

For the linear transversal filter the algorithm is a little more complicated because the stored variables are not the values of $e(iT)$ but the coefficients h_j. The commonest algorithm is given by

$$\mathbf{H}((i + 1)T) = \mathbf{H}(iT) + \mu.r(iT).\mathbf{D}(iT)$$

where $\mathbf{H}(iT)$ and $\mathbf{D}(iT)$ are the vectors of tap coefficients and data elements respectively at time iT. This is known as the stochastic gradient algorithm. If the data elements are mutually uncorrelated then the optimum values of the coefficients h_j are the first N samples of the echo impulse response. Note that, unlike the look-up table echo canceller, each coefficient is updated every iteration of the algorithm and so it converges faster.

When the echo canceller is working in the presence of the signal from the far end, that signal is superimposed on the error signal and will cause the echo canceller coefficients to fluctuate about their true values. This fluctuation will not affect the mean value of the coefficients provided the signal is not correlated with the transmitted signal, a property usually made highly likely by the use of different scramblers for the two directions of transmission.

4.12 NON-LINEARITY

As already stated the look-up table echo canceller can model a non-linear echo response. For the linear echo canceller any non-linear distortion products must be well below the level of cancellation required. Achieving the desired degree of linearity can be a problem for higher performance, longer reach systems implemented in low cost circuitry. At the same time such systems may require large echo canceller spans making the pure look-up table approach impractical. Fortunately, techniques are known that, by exploiting a knowledge of the actual non-linearity, allow the non-linear distortion to be coped with by the addition of extra echo canceller taps or by a combination of look-up tables and linear taps.

4.13 JITTER PERFORMANCE

Timing jitter is a well-known source of degradation in digital transmission systems, but in echo cancelling duplex systems it can be much more serious. If the sampling of the received signal and/or the generation of the transmitted signal is subject to jitter then the echo replica $e(iT)$ formed by the echo impulse response is sampled at slightly different phases. By having a timing recovery circuit that introduces known phase increments an echo canceller can be designed which corrects for the error.

4.14 DYNAMIC RANGE

Echo cancellers are required to suppress echoes to well below the level of the received signal from the far end. As the received signal is often highly attenuated and the hybrid circuit usually does not give a great deal of trans-hybrid loss the dynamic range requirement of the echo canceller is severe. We can write the minimum dynamic range of the echo replica signal as the product of:

peak transmit level,
decision distance attenuation,
trans-hybrid loss,
quantizer peak signal-to-rms noise ratio

The decision distance in the back-to-back case is defined as unity and so the decision distance attenuation is the pulse loss of the cable, transmit and receive filters and the linear equalizer (if used). The noise used in calculating the *S/N* ratio is the sum of any external noise arriving at the input to the receiver, any uncancelled echo and ISI and the quantization noise arising from either the DAC or ADC used in the implementation.

The dynamic range is plotted in Figure 4.18 for a number of transmission system types and assuming a *S/N* of 20 dB (corresponding to a 6 dB margin against a 1 in 10^6 error rate for Gaussian noise). It can be seen that for long loops baseband systems are essential and even then the dynamic range requirement is severe. In a digital realization the digital accuracy (1 bit per 6 dB) is acceptable, but the provision of accurate DACs and ADCs constitutes a significant practical problem.

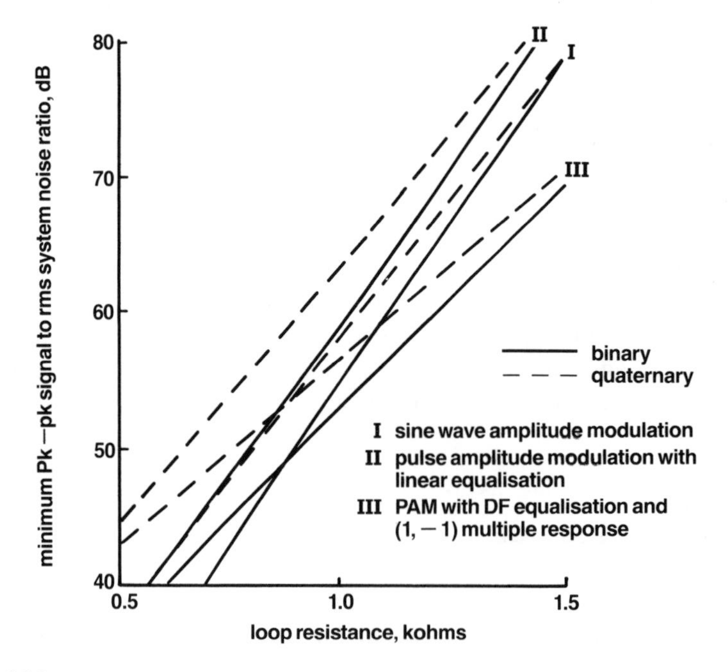

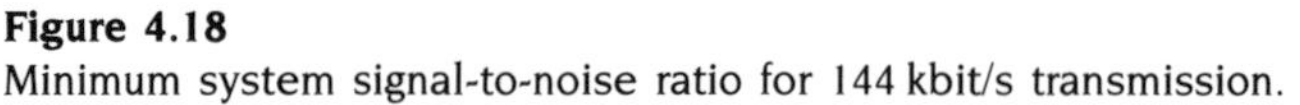

Figure 4.18
Minimum system signal-to-noise ratio for 144 kbit/s transmission.

4.15 TRANSMITTER REALIZATION

The transmitter is perhaps the easiest part of the duplex modem to realize. It does, however, in a data-driven echo cancelling system have some severe constraints imposed on its performance. The *S/N* ratio of the transmitted signal must be such that noise is acceptably below the received signal level at the input to the receiver as this cannot be removed by the echo canceller. As mentioned in Section 4.12, if a linear echo canceller is used any non-linearity in the transmitter must be insignificant. It is also advantageous if the transmit filter is constrained to be minimum phase. A non-minimum phase filter will tend to create an echo pre-cursor so that additional echo canceller taps will be necessary.

4.16 RECEIVER REALIZATION

The realization of the receiver function depends on its complexity. A receiver not requiring an adaptive equalizer may be readily realized in analogue form. The realization of an adaptive equalizer, whilst not quite as challenging as an echo canceller because of the reduced dynamic range requirements, can, nevertheless, be facilitated by digital signal processing.

4.17 LINE INTERFACING

The line interface connects the unbalanced transmitter signal to the balanced 2-wire line and converts the balanced receive signal to an unbalanced form for the input to the receiver. In addition, if the subscriber loop is required to be line powered, it must allow separation of the power from signals. A line transformer with split windings can be used for power feeding or extraction. Inductors can be used to prevent power circuitry noise from interfering with transmission. The transformer split windings must be matched to ensure that any imbalance to ground does not cause the crosstalk characteristic to worsen.

Again the line interface must not introduce any significant level of noise and must be sufficiently linear. In addition the values of inductor and transformer inductance have a marked effect on the cable pulse and echo responses and need to be carefully designed.

4.18 USA STANDARD SYSTEM

As discussed in Section 1.7, one of the perceived needs for the ISDN was to have an interface to which many terminals could be connected. In Section 1.9 the CCITT ISDN reference configuration is described and it was tacitly assumed that the customer would take responsibility for the connection at the T reference point, and that the network would be terminated by the NT1. However, in the USA it was determined legally that the NT1 would be the responsibility of the customer and that the defined interface would be at the U reference point, to provide a 'wires only' interface. For this reason a U interworking interface has to be defined. Such an interface has to define the meaning of all the bits on the interface, their temporal order and their electrical characteristics. In particular, the line code and the modulation scheme are as we have already seen of great importance in determining the performance obtained.

In August 1986 the American National Standards Institute chose to define an interface based on 2B1Q line code, PAM at 80 kbauds initially proposed by British Telecom Research Laboratories at Martlesham. The fundamental modulation process is to convert two binary digits into one quaternary element (quat). The choice was based on both performance and complexity issues. It is convenient to represent the four possible quat states as +3, +1, −1, −3 as this indicates symmetry about zero, equal spacing between states and convenient integer magnitudes.

4.19 MODULATION FORMAT

The significance of all the bits transmitted, the transmission rate and the line coding together are known as the modulation format. 2B1Q is a PAM system; because each symbol represents two binary digits to carry the 144 kbit/s of the basic rate access a modulation rate of at least 72 kbaud is required. In practice additional 'housekeeping' channels are required for framing, maintenance and control functions and hence a modulation rate of 80 kbaud was chosen. Six kbauds are allocated for framing, leaving 4 kbit/s for maintenance and control.

A frame structure was chosen based on the use of a block Synchronization Word (SW) of nine quaternary elements repeated every 1.5 ms:

$$+3,\ +3,\ -3,\ -3,\ -3,\ +3,\ -3,\ +3,\ +3.$$

It has an autocorrelation function with a peak of approximately 4 relative to the normalized rms value of a random 2B1Q signal and the off-peak values are all equal to or less than 0.2 relative to the peak. These values are a measure of the tolerance of the SW to ISI and of its efficacy at channel estimation for timing recovery respectively.

Alternative decision directed timing recovery schemes do not require a block SW and have the advantage that all the received signal is used for timing phase estimation. Therefore, they are less sensitive to noise and hence give better jitter performance than schemes which use only the block SW.

One clear advantage of the block SW is that it can be used to curtail error propagation events in a DFE based transceiver by feeding a locally generated replica of the SW through the DFE at the appropriate time each frame. In conjunction with this it can also be exploited as an alternative to jitter compensation or an analogue phase locked loop (PLL), to enable timing phase adjustment to be made with a digital PLL such that the resulting echo cancellation error does not corrupt user data.

The frame structure of the line signal consists of 120 quats (i.e. 240 bits). Eight frames together make a multiframe. The first frame of the multiframe is indicated by sending the Inverted Synchronization Word (ISW). At the end of each frame are some maintenance and control bits, including provision of an embedded maintenance and control channel, activation bit, cold start indication bit, power status bits, and far-end error check bits.

Summarizing, the frame structure is:

bits	*quats*	
1–18	1–9	Inverted Synchronization Word (frame 1) Synchronization Word (frames 2–8)
19–26	10–13	B channel 1
27–34	14–17	B channel 2
34–36	18	D channel
37–44	19–22	B
45–52	23–26	B
53–54	27	D
55–62	28–31	B
63–70	32–35	B
71–72	36	D
73–80	37–40	B
81–88	41–44	B
89–90	45	D
91–98	46–49	B
99–106	50–53	B
107–108	54	D
109–116	55–58	B
117–124	59–62	B
125–126	63	D
127–134	64–67	B

bits	*quats*	
135–142	68–71	B
143–144	72	D
145–152	73–76	B
153–160	77–80	B
161–162	81	D
163–170	82–85	B
171–178	86–89	B
179–180	90	D
181–188	91–94	B
189–196	95–98	B
197–198	99	D
199–206	100–103	B
207–214	104–107	B
215–216	108	D
217–224	109–112	B
225–232	113–116	B
233–234	117	D
235–240	118–120	Maintenance and control

The frame repeats every 1.5 ms.

As mentioned in Section 4.11, for proper operation of the echo cancellers there must be no correlation between the signals in opposite directions and this is achieved by inserting scramblers. For the USA system these consist of shift registers with modulo-two additions as shown in Figure 4.19. The synchronization words are not scrambled, but are offset by half a frame in each direction of transmission to avoid correlation.

4.20 TRANSMIT SIGNAL SPECIFICATION

It is important that transceivers of different manufacture can interwork without any significant loss of performance. To achieve this, whilst still allowing a reasonable spread of component tolerances, the following features of the line signal are specified.

Transmit signal power

To minimize the effect of impulsive noise it is desirable to have as high a signal level as possible. The signal level has no effect on crosstalk between ISDN systems as the interfering level and the interfered signal level will be

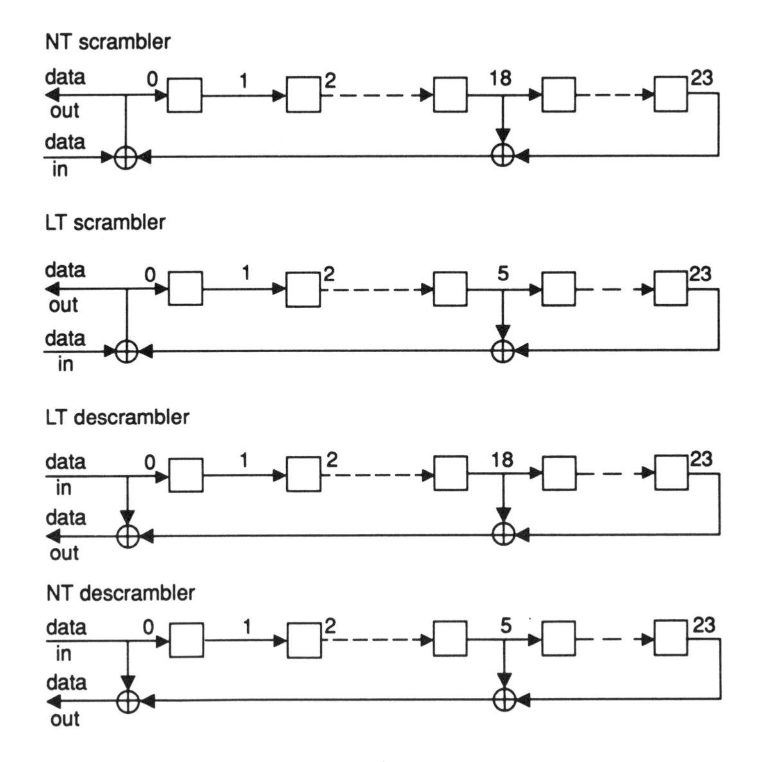

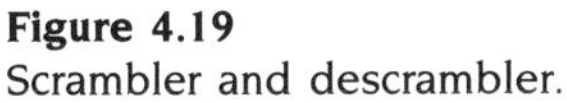

Figure 4.19
Scrambler and descrambler.

similarly affected. The signal level is limited by two factors. First the need to avoid crosstalk interference to other non-ISDN services in the cable. Secondly to keep within the range of practical VLSI circuit output powers. The compromise of 13.5 ± 0.5 dBm into 135 Ω is specified.

Pulse shape

The shape of the pulse implies a power spectrum and peak voltage which are chosen to enable the ISDN to coexist with other services. A pulse mask is defined which controls the variation between transceivers of different manufacture. The peak signal level is 2.5 V representing a +3 quat.

Non-linearity

It is possible for output pulses to lie within the pulse mask but still to be significantly different from a linear function of the generating quat. As discussed earlier, it is possible for equalizers and echo cancellers to cope with non-linear distortion but only at the expense of greater complexity and other degradations. To put a bound on the level of this non-linearity it is specified that the non-linear portion of the line signal power must be at least 36 dB below the signal power.

Frequency offset and jitter

Although the nominal modulation rate is 80 kbaud there are two effects which change this. First the clock to which the network is tied may be in error. In practice, although figures of 5 p.p.m. tolerance may be specified, the actual frequency offset is normally considerably less than this. The other source is jitter which is short-term variation of the timing waveform which arises from many sources, including the imperfect extraction of timing information from digital line systems.

The jitter is described in terms of its magnitude measured in the width of a symbol, known as a 'unit interval' (UI). The magnitude of the tolerable jitter varies with its frequency. Below 0.5 Hz it can be 0.3 UI pk–pk. Above 0.5 Hz the permitted jitter decreases inversely with frequency to 0.008 UI pk–pk at 19 Hz. Typically the free-running clocks in the NT1 are within 50 p.p.m. of the standard frequency. The low frequency jitter is approximately equivalent to a 6 p.p.m. frequency offset. Together with the possible network error this means that the NT1 must cope with a possible 61 p.p.m. offset (50 + 6 + 5).

At the customer's end the pulses generated by the NT1 will derive their timing from those received from the exchange. As clock recovery is not perfect, additional jitter may be added, which must be tracked by the exchange line termination. The additional jitter must be less than 0.04 UI pk–pk and 0.01 UI rms when measured with an 80 Hz high-pass filter and must be less than 0.05 UK pk–pk and 0.015 UI rms when measured in a band between 1 and 40 Hz; in all cases the filters roll-off at 6 dB/octave. The variation with time between the output pulse phase and input pulse phase must not exceed 0.1 UI.

Start-up procedure

As discussed in Section 4.11, the echo canceller is self-adapting to the signal presented to it, but the rate of convergence of the process can very much depend on its detailed design. Similar considerations apply to convergence of DFEs and clock extraction circuits. Steps can be taken to minimize this start-up time including the use of special training sequences initially to which the echo canceller and DFE can easily converge, and the storing from one call to another of information so that only a small update is required to take into account changes which may have taken place (e.g. cable temperature). The start-up procedure is defined flexibly and simply so that different transceiver designs are possible and so that the start-up time may be used in different ways according to the particular design. For a cold start (i.e. when the equipment is first used) 5 s are allowed for the NTI and 10 s for the exchange. As these are additive a total of 15 s may elapse. A warm start is allowed a total of 300 ms. In this case the equipment will have been used previously and echo canceller coefficients, DFE coefficients and exchange timing phase may have been stored from the previous use and may be used to facilitate a faster transceiver start-up.

4.21 LOCAL NETWORK TRANSMISSION OUTSIDE USA

As mentioned in Section 4.18, in the USA legal constraints have been imposed which require the customer interface to be at the U reference point. This was not originally foreseen by CCITT, who assumed that the customer interface would be at the T reference point and that the NT1 would be regarded as part of the network. This is mainly because the interface at the T reference point is more easily defined and not dependent on the implementation and characteristics of the particular administration's local network. In the future the local network copper pairs may well be replaced by optical fibres or radio. The inclusion of the NT1 in the network shields the customer from the technological change elsewhere. For this reason there is no international standard for the U reference point. On copper pairs, Germany has used an echo cancelling system which uses a line code which converts four binary digits into three ternary digits (4B3T). The UK has used a 3B2T system but is moving to 2B1Q. Other countries are using AMI line code; however, this is very restricted on range.

Nevertheless, with the widespread availability of 2B1Q devices to meet the USA market, it is likely that these will also be used by other administrations where service is still provided on copper pairs.

REFERENCE

American National Standards Institute: ANSI T1.601–1988. ISDN Basic Access Interface for use on Metallic Loops for application on the Network side of the NT (Layer 1 specification).

QUESTIONS

1 The transmission system chosen for the local network is designed to perform adequately on about 99% of local network pairs. If satisfactory performances were required on only 50% of pairs, how many ISDN B channels could be carried on each pair?

2 Why is a simple hybrid arrangement inadequate for digital duplex transmission in the local network?

3 Echo cancellers have two basic structures—look-up table and transversal filter. What are the advantages and disadvantages of each?

4 Why are scramblers used on local network transmission systems?

Chapter 5

The Basic Rate ISDN Customer's Interface

It has already been mentioned in Section 1.7 that CCITT decided to offer two 64 kbit/s B channels as the basic access to the customer plus a D signalling channel of 16 kbit/s. It is called a 'basic' access to distinguish it from other formats such as 30B + D channels in Europe and 23B + D channels in North America which are known as 'primary rate' accesses, and from reading Section 3.1 will be recognized as being of the same form as the inter-exchange transmission systems. Primary rate access is discussed further in Chapter 6.

The basic rate access is intended to be equivalent in ISDN terms to the simple telephone in analogue terms. The connection of a multiplicity of analogue terminals (e.g. telephones, modems) is achieved by simply connecting them in parallel. The requirement for the ISDN interface is also to be able to simply connect terminals in parallel with the additional feature that terminals may be individually addressed by type or other identifier.

It will be noted that the detailed description of the signalling protocols is often given in CCITT Standards using 'Specification and Description Language' (SDL). The casual user will immediately get an intuitive understanding of what it is conveying. However, a fuller understanding will give a greater insight into the processes involved and so an appendix describing the language is included at the end of the book.

5.1 THE CUSTOMER'S INSTALLATION

John Hovell

The CCITT reference configuration for basic rate access described in Sections 1.7 and 1.9 is represented pictorially in Figure 5.1.

The two-wire transmission line from the local network is terminated by the NT1. Although this NT1 is located on the customer's premises it is on the network side of the CCITT defined user–network interface and hence the responsibility of the network operator. The exception to this is in the

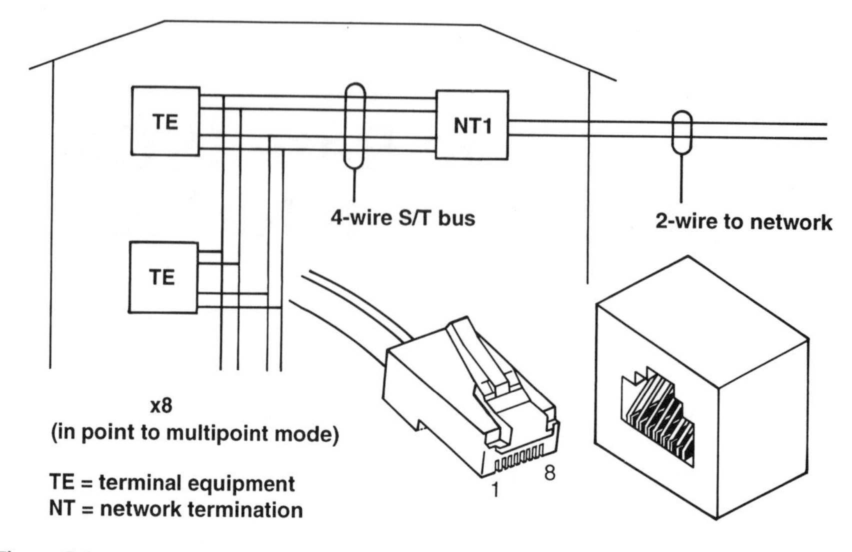

Figure 5.1
User–network coniguration and connector.

USA where the user–network interface is at the network side of the NT1 as explained in Section 4.18.

Into this NT1 is connected a four-wire bus consisting of a transmit and receive pair often known as the 'S' bus because it is at the CCITT S reference point. This bus is designed to operate over normal twisted pair internal cable as may be provided for traditional analogue sources. The bus can operate in two modes, point-to-point or point-to-multipoint. In the point-to-point mode one Terminal Equipment (TE) is connected at the end of up to a maximum of about 1 km of cable; the actual limit is an attenuation of 6 dB at 96 kHz. In the point-to-multipoint mode up to eight terminals can be connected in parallel anywhere along the bus, but the bus length is now limited to about 200 m by timing constraints. The terminals are high impedance (2500 Ω) on both their transmit and receive ports and so do not load the bus. The bus is terminated by 100 Ω resistors at both the NT1 and the distant end, on both the transmit and receive pairs. Connections are made to the bus via 8-pin plugs and sockets to the International Standards Organization (ISO) Specification 8877. These are a development of the standard USA 6-pin analogue telephone connector.

Over this bus passes the two transparent 64 kbit/s B channels, the 16 kbit/s D signalling channel and other bits used for miscellaneous purposes such as frame synchronization and collision avoidance. This contributes another 48 kbit/s making a total of 192 kbit/s.

The B channels contain the user data which is switched by the network to provide an end-to-end transmission service. A B channel path is establis-hed by signalling messages in the D channel.

In a multi-terminal situation all terminals have access to the D channel by the use of an access procedure but each B channel is allocated to a particular terminal at the time of call set-up and is not capable of being shared between terminals.

5.1.1 Power feeding

Power is provided across the user–network interface to provide a basic telephone service in the event of a local mains failure. The physical arrangement for power feeding is shown in Figure 5.2. The centre of the diagram shows a basic four-wire bus consisting of two twisted pairs. A power source 1 is available within the NT which can supply, via the centre tapped transformers (known as a phantom feed), at a nominal 40 V, up to 1 W of power. The power under these normal arrangements will probably be derived at the NT1 from a local mains source. Under mains failure conditions the power will be limited to 420 mW, and to indicate this state to the terminals the polarity will be reversed. In this case the power will be derived from the network, and is only intended to power a single digital telephone for emergency use. Note that there are an additional two optional pairs allocated for alternative power feeding arrangements, but it is not clear the extent to which these will be used.

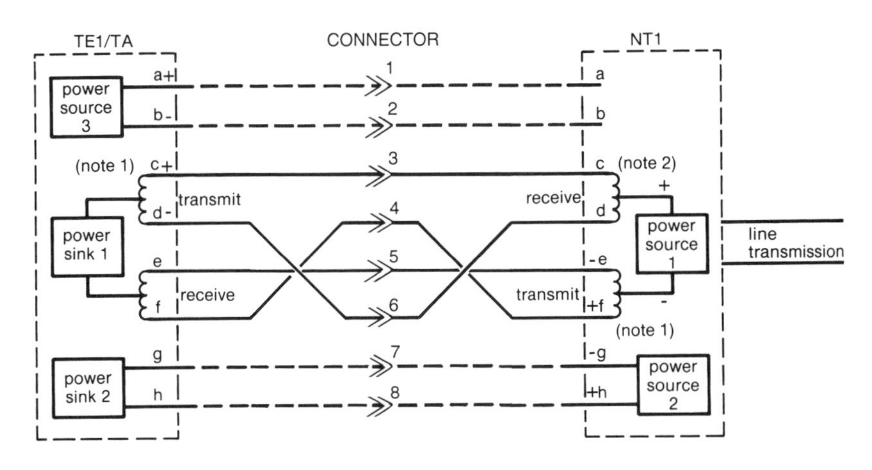

Figure 5.2
Reference configuration for signal transmission and power feeding in normal operating mode.

5.1.2 Layer 1 of the user–network interface

Layer 1 is responsible for transferring information between terminals and the NT1. The transmission medium on which it must achieve this is not perfect and errors may well be introduced. For the D channel signalling (the C-plane) between terminals and the network it is Layer 2's responsibility to handle any error detection and correction. For user information on the B channels (the U-plane) the network offers no error control although it must be stressed that the error rate is expected to be very low.

5.1.3 The functions of Layer 1

In order to provide this transport service Layer 1 must support the functions outlined below.

B channels

Layer 1 must support for each direction of transmission, two independent 64 kbit/s B channels. The B channels contain user data which is switched by the network to provide the end-to-end transmission source. Note there is no error correction provided by the network on these channels.

D channels

Layer 1 must support for each direction of transmission, a 16 kbit/s channel for the signalling information. In some networks user packet data may also be supported on the D channel.

D channel access procedure

This procedure ensures that in the case of two or more terminals on a point-to-multipoint configuration attempting to access the D channel simultaneously, one terminal will always successfully complete the transmission of information. This procedure is described in more detail later.

Deactivation

Deactivation permits the TE and NT equipments to be placed in a low power mode when no calls are in progress; activation restores the equipment to its normal power mode. A low power mode is required for two purposes. First to save power. This is significant from the network point of view as the NT1 is line powered. Secondly, deactivation is also provided to reduce electromagnetic radiation and hence crosstalk interference.

The interface may be activated from either the terminal or network side, but deactivated only from the network side because of the NT1's multi-terminal capability. This procedure is described in more detail later.

5.1.4 Binary organization of Layer 1 frame

The structures of the Layer 1 frames across the interface are different in each direction of transmission. Both structures are shown in Figure 5.3.

A frame is 48 bits long and lasts 250 μs. The bit-rate is therefore 192 kbit/s and each bit approximately 5.2 μs long. Figure 5.3 shows that there is a 2-bit offset between transmit and receive frames. This is the delay between a frame start at the receiver of a terminal and the frame start of the transmitted signal.

Figure 5.3 also illustrates that the line code used is Alternate Mark Inversion (AMI); a logical 1 is transmitted as zero voltage and a logical 0 as a positive or negative pulse. Note that this convention is the inverse of that used on line transmission systems (see Section 3.2). The nominal pulse amplitude is 750 mV.

A frame contains several L bits; these are balance bits to prevent a build-up of d.c. on the line. The frames are split into balanced blocks as represented by dots on the diagram. For the direction TE to NT, where each B channel may come from a different terminal, each terminal's output contains an L bit to form a balanced block.

Examining the frame in the NT to TE direction, the first bits of the frame are the F/L pair, which is used in the frame alignment procedure. The start of a new frame is signalled by the F/L pair violating the AMI rules. Obviously once a violation has occurred, to restore correct polarity before the next frame, there must be a second violation. This takes place with the first mark after the F/L pair. The FA bit ensures this second violation occurs should there not be a mark in the B1, B2, D, E, or A channels. The E channel is an echo channel in which D-channel bits arriving at the NT1 are echoed back to the TEs. There is a 10-bit offset between the D channel leaving a terminal, travelling to the NT and being echoed back in the E channel.

The A bit is used in the activation procedure to indicate to the terminals that the system is in synchronization. Next is a byte of the B2 channel, a bit of the E channel and a bit of the D channel, followed by a M bit. This is used for multiframing, a facility not provided in Europe, but which may be

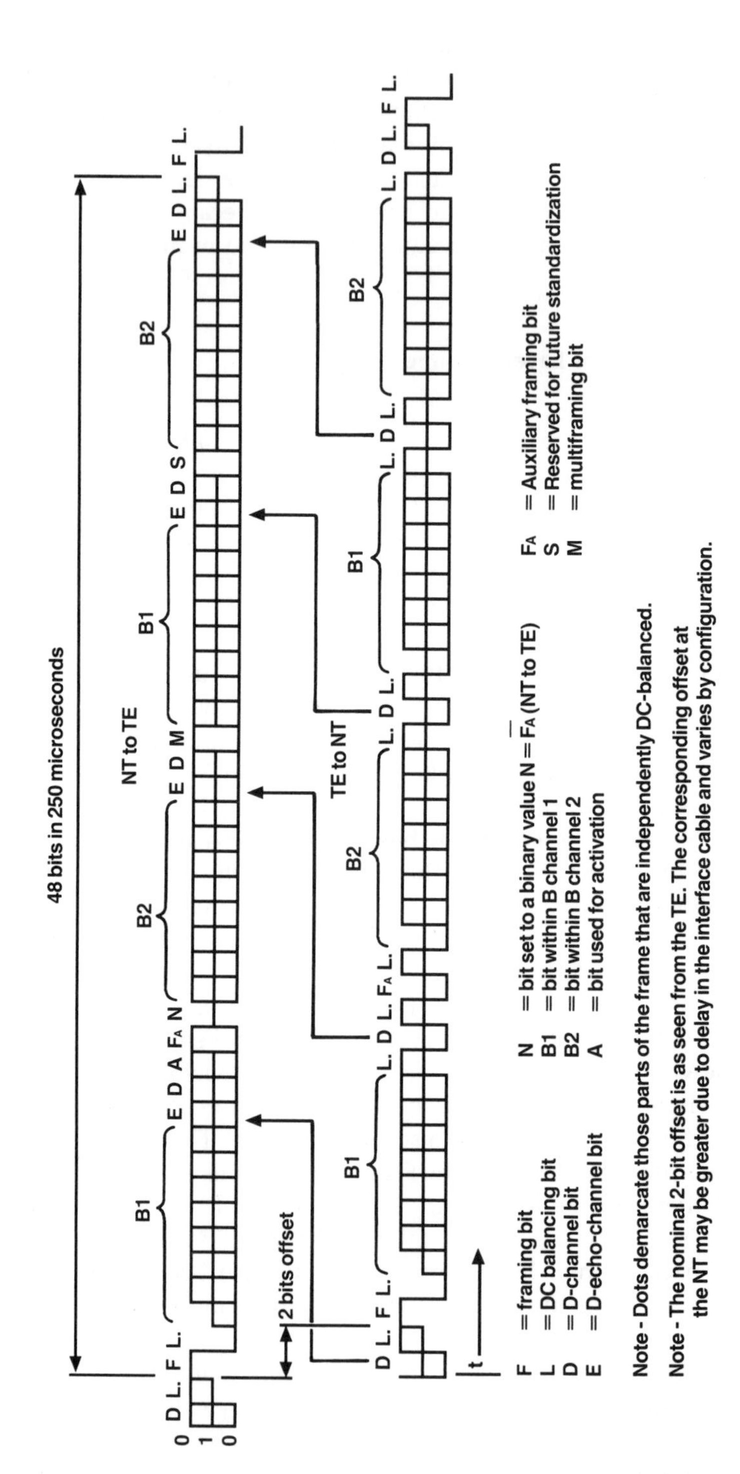

Figure 5.3
Frame structure.

used elsewhere. The M bit identifies some FA bits which can be stolen to provide a management channel.

The B1, B2, D and E channels are then repeated along with the S bit which is a spare bit.

5.1.5 Layer 1 D channel contention procedure

This procedure ensures that, even in the case of two or more terminals attempting to access the D channel simultaneously, one terminal will always successfully complete the transmission of information. The procedure relies on the fact that the information to be transmitted consists of Layer 2 frames delimited by flags consisting of the binary pattern 01111110. Layer 2 applies a zero bit insertion algorithm to prevent flag imitation by a Layer 2 frame. The interframe time fill conists of binary 1s which are represented by zero voltage. The zero volt line signal is generated by the TE transmitter going high impedance. This means a binary 0 from a parallel terminal will overwrite a binary 1. Detection of a collision is done by the terminal monitoring the E channel (the D channel echoed from the NT).

To access the D channel a terminal looks for the interframe time fill by counting the number of consecutive binary 1s in the D channel. Should a binary 0 be received the count is reset. When the number of consecutive 1s reaches a predetermined value (which is greater than the number of consecutive 1s possible in a frame because of the zero bit insertion algorithm) the counter is reset and the terminal may access the D channel. When a terminal has just completed transmitting a frame the value of the count needed to be reached before another frame may be transmitted is incremented by 1. This gives other terminals a chance to access the channel. Hence an access and priority mechanism is established.

There is still the possibility of a collision between two terminals of the same priority. This is detected and resolved by each terminal comparing its last transmitted bit with the next E bit. If they are the same the terminals continue to transmit. If, however, they are different the terminal detecting the difference ceases transmission immediately and returns to the D channel monitoring state leaving the other terminal to continue transmission.

5.1.6 Layer 1 activation/deactivation procedure

This procedure permits activation of the interface from both the terminal and network side, but deactivation only from the network side. This is because of the multi-terminal capability of the interface. Activation and deactivation information is conveyed across the interface by the use of line signals called 'Info signals'.

Info 0 is the absence of any line signal; this is the idle state with neither terminals nor the NT working.

Info 1 is flags transmitted from a terminal to the NT to request activation. Note this signal is not synchronized to the network.

Info 2 is transmitted from the NT to the TEs to request their activation or to indicate that the NT has activated as a response to receiving an Info 1. An Info 2 consists of Layer 1 frames with a high density of binary zeros in the data channels which permits fast synchronization of the terminals.

Info 3 and Info 4 are frames containing operational data transmitted from the TE and NT respectively.

The principal activation sequence is commenced when a terminal transmits an Info 1. The NT activates the local transmission system which indicates to the exchange that the customer is activating. The NT1 responds to the terminals with an Info 2 to which the TEs synchronize. The TEs respond with an Info 3 containing operational data and the NT is then in a position to send Info 4 frames. Note that all terminals activate in parallel; it is not possible to have just one terminal activated in a multi-terminal configuration. The network activates the bus by the exchange activating the local network transmission system. Deactivation occurs when the exchange deactivates the local network transmission system.

5.1.7 Layer 1 physical implementation

Semiconductor manufacturers are introducing devices to perform the functions described above. This is often in the form of two chips as illustrated in Figure 5.4, one providing the transmission functions and the other digital logic. The first chip contains the transmitters and receivers with associated timing extraction. The other chip contains a frame formatter, channel multiplexing and demultiplexing and a microprocessor interface for control of both chips. This second chip often also contains the lower layer functions of the Layer 2 HDLC processing for the D channel. To complete the circuitry a line transformer is used to provide a balanced line signal and d.c. isolation. To protect the silicon from induced transients on the bus, protection diodes are required and, ideally, some series impedance to dissipate the power.

Also becoming available are d.c.–d.c. power converters specifically designed for this ISDN application. The NT power controllers provide the

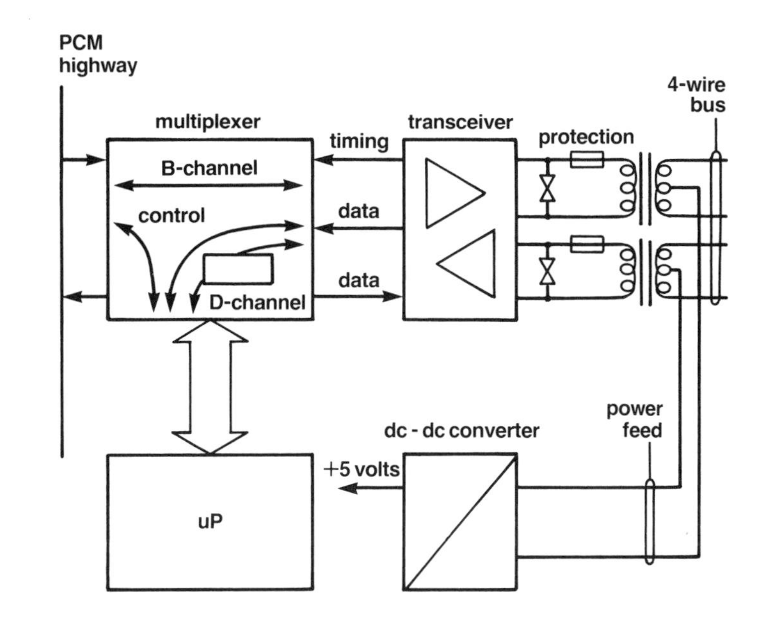

Figure 5.4
Typical Layer 1 implementation—terminal side.

40 V phantom line feed from a 50 V source. The TE power controllers convert the received line voltage into 5 V for driving the TE circuitry.

5.2 LAYER 2

Kevin Woollard

The Layer 2 Recommendation describes the high level data link (HDLC) procedures commonly referred to as the Link Access Procedure for a D channel or LAP D. The objective of Layer 2 is to provide a secure, error-free connection between two end-points connected by a physical medium. Layer 3 call control information is carried in the information elements of Layer 2 frames and it must be delivered in sequence and without error. Layer 2 also has responsibility for detecting and retransmitting lost frames.

LAP D was based originally on Lap B of the X.25 Layer 2 Recommendation. However, certain features of LAP D give it significant advantages. The most striking difference is the possibility of frame multiplexing by having separate addresses at Layer 2 allowing many LAPs to exist on the same physical connection. It is this feature that allows up to eight terminals to share the signalling channel in the passive bus arrangement.

Each Layer 2 connection is a separate LAP and the termination points for the LAPs are within the terminals at one end and at the periphery of the

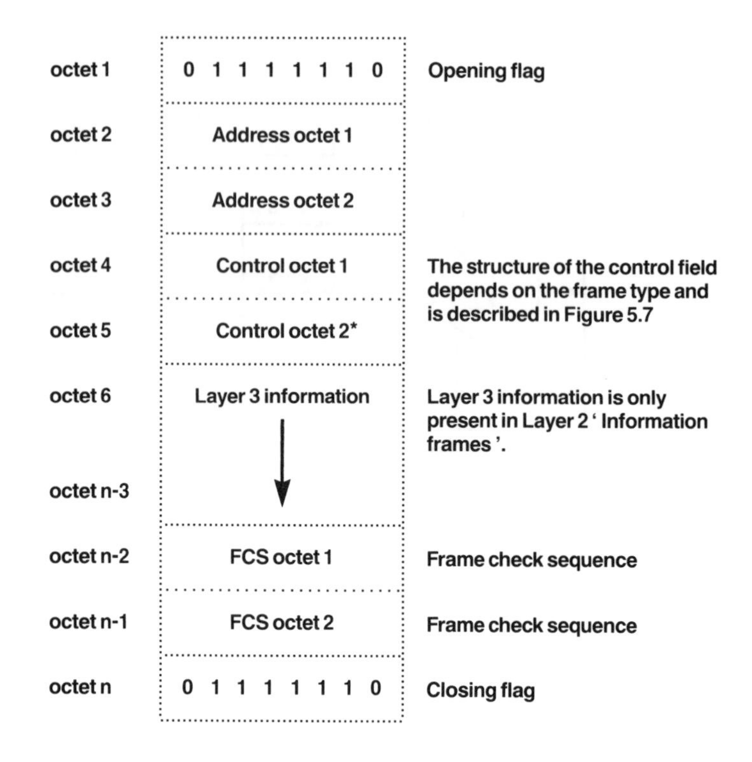

Figure 5.5
Layer 2 frame structure.

exchange at the other. Layer 2 operates as a series of frame exchanges between the two communicating, or peer, entities. The frames consist of a sequence of 8-bit elements and the elements in the sequence define their meaning as shown in Figure 5.5. A fixed pattern, called a flag, is used to indicate both the beginning and end of a frame. Two octets are needed for the Layer 2 address and carry a Service Access Point Identifier (SAPI), a terminal identifier (TEI) and a command/response bit. The control field is one or two octets depending on the frame type and carries information that identifies the frame and the Layer 2 sequence numbers used for link control. The information element is only present in frames that carry Layer 3 information and the Frame Check Sequence (FCS) is used for error detection. A detailed breakdown of the individual elements is given in Figures 5.6 to 5.8.

What cannot be shown in the diagrams is the procedure to avoid imitation of the flag by the data octets. This is achieved by examining the serial stream between flags and inserting an extra 0 after any run of five 1 bits. The receiving Layer 2 entity discards a 0 bit if it is preceded by five 1s.

The two octets of the address field are coded as shown:

8	7	6	5	4	3	2	1	
S A P I						C/R	EA0	octet 2
T E I							EA1	octet 3

SAPI — Service access point identifier
TEI — Terminal endpoint identifier
C/R — Command/response bit
EA0 — Extended address bit 0
EA1 — Extended address bit 1

Figure 5.6
The Layer 2 address frame.

8	7	6	5	4	3	2	1	
N(S)							0	octet 4
N(R)							P	octet 5

Control field format for information frames

8	7	6	5	4	3	2	1	
X	X	X	X	S	S	0	1	octet 4
N(R)							P	octet 5

Control field format for supervisory frames

8	7	6	5	4	3	2	1	
M	M	M	P/F	M	M	1	1	octet 4

Control field format for unnumbered frames

N(S) Transmitter send sequence number
N(R) Transmitter receive sequence number
S Supervisory function bit
M Modifier function bit
P/F Poll bit when issued as a command, final bit when issued as a response
X Reserved and set to 0

Figure 5.7
Control field formats. The controlled field may be 1 or 2 octets in length depending on the frame type. The detail of the control field formats are given in the figure.

COMMAND	RESPONSE	ENCODING								
		8	7	6	5	4	3	2	1	
INFORMATION FRAMES										
I FRAME	—	N(S)							0	4
		N(R)							P	5
SUPERVISORY FRAMES										
RR	RR	0	0	0	0	0	0	0	1	4
		N(R)							P/F	5
RNR	RNR	0	0	0	0	0	1	0	1	4
		N(R)							P/F	5
REJ	REJ	0	0	0	0	1	0	0	1	4
		N(R)							P/F	5
UNNUMBERED FRAMES										
SABME	—	0	1	1	P	1	1	1	1	4
—	DM	0	0	0	F	1	1	1	1	4
UI	—	0	0	0	P	0	0	1	1	4
DISC	—	0	1	0	P	0	0	1	1	4
—	UA	0	1	1	F	0	0	1	1	4
—	FRMR	1	0	0	F	0	1	1	1	4

I FRAME	Information frame	DISC	Disconnect
RR	Receiver ready	UA	Unnumbered acknowledgement
RNR	Receiver not ready	FRMR	Frame reject
REJ	Reject	DM	Disconnect mode
SABME	Set asynchronous balanced mode extended	UI	Unnumbered information

Figure 5.8
Layer 2 frame control field codings. Three Layer 2 frame types have been identified in Figure 5.7. By assigning different combinations to the S and M bits many frames may be defined. The figure shows the control fields of all the Layer 2 frames and indicates whether a frame exists as a command, response or both.

5.2.1 Layer 2 addressing

Layer 2 multiplexing is achieved by employing a separate Layer 2 address for each LAP in the system. To carry the LAP identity the address is two octets long and identifies the intended receiver of a command frame and the transmitter of a response frame. The address has only local significance and is known only to the two end-points using the LAP. No use can be made of the address by the network for routing purposes and no information about its value will be held outside the Layer 2 entity.

The Layer 2 address is constructed as shown in Figure 5.6. The details of each element will be described in the next section but it is useful to introduce the concepts associated with the SAPI and TEI fields.

The Service Access Point Identifier (SAPI) is used to identify the service that the signalling frame is intended for. An extension of the use of the D channel is to use it for access to a packet service as well as for signalling. Consider the case of digital telephones sharing a passive bus with packet terminals. The two terminal types will be accessing different services and possibly different networks. It is possible to identify the service being invoked by using a different SAPI for each service. This gives the network the option of handling the signalling associated with different services in separate modules. In a multi-network ISDN it allows Layer 2 routing to the appropriate network. The value of the SAPI is fixed for a given service.

The Terminal Endpoint Identifier (TEI) takes a range of values that are associated with terminals on the customer's line. In the simplest case each terminal will have a single unique TEI value. The combination of TEI and SAPI identify the LAP and provide a unique Layer 2 address. A terminal will use its Layer 2 address in all transmitted frames and only frames received carrying the correct address will be processed.

In practice a frame originating from telephony call control has a SAPI that identifies the frame as 'telephony' and all telephone equipment will examine this frame. Only the terminal whose TEI agrees with that carried by the frame will pass it to the Layer 2 and Layer 3 entities for processing.

As it is important that no two TEIs are the same, the network has a special TEI management entity which allocates TEI on request and ensures their correct use. The values that TEIs can take fall into the ranges:

0–63 non-automatic assignment TEIs
64–126 automatic assignment TEIs
127 global TEI

Non-automatic TEIs are selected by the user; their allocation is the responsibility of the user. Automatic TEIs are selected by the network; their allocation is the responsibility of the network. The mechanism is discussed in Section 5.2.5. The global TEI is permanently allocated and is often referred to as the broadcast TEI.

Terminals which use TEIs in the range 0–63 need not negotiate with the network before establishing a Layer 2 connection. Nevertheless the rule that all TEIs on the customer's line must be different still applies and it is the responsibility of the user to ensure that he does not connect two terminals using the same non-automatic TEI.

Terminals which use TEIs in the range 64–126 cannot establish a Layer 2 connection until they have requested a TEI from the network. In this case it is the responsibility of the network not to allocate the same TEI more than once to any given time.

The global TEI is used to broadcast information to all terminals within a given SAPI; for example, a broadcast message to all telephones, offering an incoming telephony call.

5.2.2 Layer 2 operation

The function of Layer 2 is to deliver Layer 3 frames, across a Layer 1 interface, error free and in sequence. It is necessary for a Layer 2 entity to interface to both Layer 1 and Layer 3. To highlight the operation of Layer 2 we will consider the operation of a terminal as it attempts to signal with the network.

It is the action of attempting to establish a call that causes the protocol exchange between terminal and network. If there has been no previous communication it is necessary to activate the interface in a controlled way. A request for service from the customer results in Layer 3 requesting a service from Layer 2. Layer 2 cannot offer a service unless Layer 1 is available and so the appropriate request is made to Layer 1. Layer 1 then initiates its start-up procedure and the physical link becomes available for Layer 2 frames. Before Layer 2 is ready to offer its services to Layer 3 it must initiate the Layer 2 start-up procedure known as 'establishing a LAP'.

LAP establishment is achieved by the exchange of Layer 2 frames between the Layer 2 handler in the terminal and the corresponding Layer 2 handler in the network (the peer entity). The purpose of this exchange is to align the state variables that will be used to ensure the correct sequencing of information frames. Before the LAP has been established the only frames that may be transmitted are unnumbered frames. The establishment procedure requires one end-point to transmit a Set Asynchronous Balanced Mode Extended (SABME) and the far end to acknowledge it with an Unnumbered Acknowledgement (UA).

Once the LAP is established Layer 2 is able to carry the Layer 3 information and is said to be the 'multiple frame established state'. In this state Layer 2 operates its frame protection mechanisms.

Figure 5.9 shows a normal Layer 2 frame exchange. Once established the LAP operates an acknowledged service in which every information frame (I frame) must be responded to by the peer entity. The most basic response is the Receiver Ready (RR) response frame. Figure 5.9 shows the LAP establishment and the subsequent I frame RR exchanges. The number of I frames allowed to be outstanding without an acknowledgement is defined as the window size and can vary between 1 and 127. For telephony

```
terminal                network

SABME ---------->
<-------------- UA
IFRAME --------->
<-------------- RR
IFRAME --------->
<-------------- RR
<--------- IFRAME
RR ------------->
```

Figure 5.9
A normal Layer 2 frame exchange.

signalling applications the window size is 1 and after transmitting an I frame the Layer 2 entity will await a response from the corresponding peer entity before attempting to transmit the next I frame. Providing there are no errors all that would be observed on the bus would be the exchange of I frames and RR responses. However Layer 2 is able to maintain the correct flow of information, in the face of many different error types.

5.2.3 Layer 2 error control

It is unlikely that a frame will disappear completely but it is possible for frames to be corrupted by noise at Layer 1. Corrupted frames will be received with invalid Frame Check Sequence (FCS) values and consequently discarded.

The frame check sequence is generated by dividing the bit sequence starting at the address up to (but not including) the start of the frame check sequence by the generator polynomial $X^{16} + X^{12} + X^5 + 1$. In practical terms this is done by a shift register as shown in Figure 5.10. All registers are preset to 1 initially. At the end of the protected bits the shift register contains the remainder from the division. The 1's complement of the remainder is the FCS. At the receiver the same process is gone through, but this time the FCS is included in the division process. In the absence of transmission errors the remainder should always be 0001 1101 0000 1111.

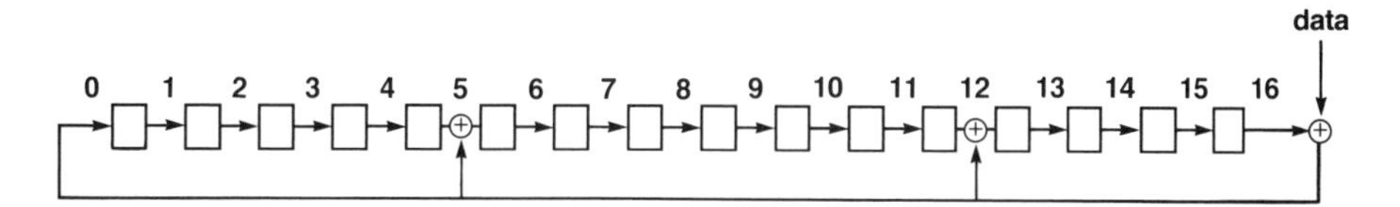

Figure 5.10
Generation of frame check sequence.

The method for recovering from a lost frame is based on the expiry of a timer. A timer is started every time a command frame is transmitted and is stopped when the appropriate response is received. This single timer is thus able to protect both the command and response frame as the loss of either will cause it to expire.

When the timer expires it is not possible to tell which of the two frames has been lost and the action taken is the same in both cases. Upon the timer expiring, the Layer 2 transmits a command frame with the poll bit set. This frame forces the peer to transmit a response that indicates the value held by the state variables. It is possible to tell from the value carried by the response frame whether or not the original frame was received. If the first frame was received, the solicited response frame will be the same as the lost response frame and is an acceptable acknowledgement. If, however, the original frame was lost, the solicited response will not be an appropriate acknowledgement and the Layer 2 entity will know that a retransmission is required.

It is possible for the same frame to be lost more than once and Layer 2 will retransmit the frame three times. If after three retransmissions of the frame the correct response has not been received, Layer 2 will assume that the connection has failed and will attempt to re-establish the LAP. Figure 5.11 illustrates a Layer 2 sequence recovering from two frame losses.

Another possible protocol error is the arrival of an I frame with an invalid send sequence number N(S). This error is more likely to occur when the LAP is operating with a window size greater than one. If, for example, the third frame in a sequence of four is lost the receiving Layer 2 entity will know that a frame has been lost from the discontinuity in the sequence numbers. The Layer 2 must not acknowledge the fourth frame as this will

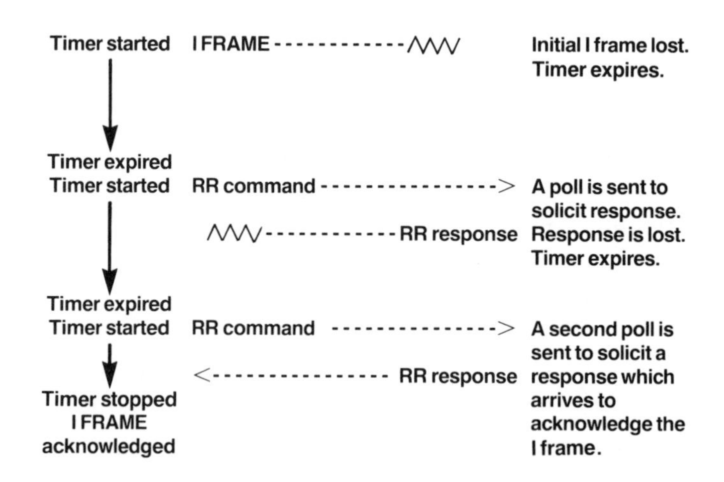

Figure 5.11
Layer 2 recovery.

imply acknowledgement of the lost third frame. The correction operation for a Reject (REJ) frame to be returned with the receive sequence number N(R) equal to N(S) + 1, where N(S) is the send variable of the last correctly received I frame, is in this case I frame two. This does two things; first it acknowledges all the outstanding I frames up to and including the second I frame, and secondly it causes the sending end to retransmit all outstanding I frames starting with the lost third frame.

The receipt of a frame with an out of sequence, or invalid, N(R) does not indicate a frame loss and cannot be corrected by retransmissions. It is necessary in this case to re-establish the LAP to realign the state variables at each end of the link.

The Receiver Not Ready (RNR) frame is used to inhibit the peer Layer 2 from transmitting I frames. The reasons for wanting to do this are not detailed in the specification but it is possible to imagine a situation where Layer 3 is only one of many functions to be serviced by a microprocessor and a job of higher priority requires that no Layer 3 processing is performed.

Another frame specified in Layer 2 is the FRaMe Reject frame (FRMR). This frame may be received by a Layer 2 entity but may not be transmitted. It is included in the recommendation to preserve alignment between LAP D and LAP B. After the detection of a frame reject condition the data link is reset.

5.2.4 Disconnecting the LAP

After Layer 3 has released the call it informs Layer 2 that it no longer requires a service. Layer 2 then performs its own disconnection procedures so that ultimately Layer 1 can disconnect and the transmission systems associated with the local line and the customer's bus can be deactivated.

Layer 2 disconnection is achieved when the frames disconnect (DISC) and UA are exchanged between peers. At this point the LAP can no longer support the exchange of I frames and supervisory frames.

The last frame type to be considered is the Disconnect Mode (DM) frame. This frame is an unnumbered acknowledgement and may be used in the same way as a UA frame. It is used as a response to a SABME if the Layer 2 entity is unable to establish the LAP, and a response to a DISC if the Layer 2 entity has already disconnected the LAP.

5.2.5 TEI allocation

Because each terminal must operate using a unique TEI, procedures have been defined in a Layer 2 management entity to control their use. The TEI

```
Terminal                                          Network

    UI TEI REQUEST ------------------------>
    (REF No. X)
    <------------------------ UI TEI ALLOCATE
                              (REF No. X)
    SABME ------------------------------->
    <---------------------------------- UA
    I FRAME ----------------------------->
    <---------------------------------- RR

                      etc.
```

Figure 5.12
TEI allocation.

manager has the ability to allocate, remove, check and verify TEIs that are in use on the customer's bus. As the management entity is a separate service point all messages associated with TEI management are transmitted with a management SAPI.

TEI management procedures must operate regardless of the Layer 2 state and so the unnumbered information (UI) frame is used for all management messages. The UI frames have no Layer 2 response and protection of the frame content is achieved by multiple transmissions of the frame.

In order to communicate with terminals which have not yet been allocated TEI, a global TEI is used. All management frames are transmitted on a broadcast TEI which is associated with a LAP that is always available. All terminals can transmit and receive on the broadcast TEI as well as their own unique TEI. All terminals on the customer's line will process all management frames. To ensure that only one terminal acts upon a frame a unique reference number is passed between the terminal and network. This reference number is contained within an element in the UI frame and is either a number randomly generated by the terminal, or 0 is the TEI of the terminal, depending on the exact situation. Figure 5.12 shows the frame exchange required for a terminal to be allocated a TEI and establish its data link connection.

5.3 Layer 3

Derek Rumsey

This layer effects the establishment and control of connections. It is carried in Layer 2 frames as can be seen in Figure 5.5. The messages are passed between the terminals and the exchange and vice versa. The first procedures defined relate to the control of circuit-switched calls. However later

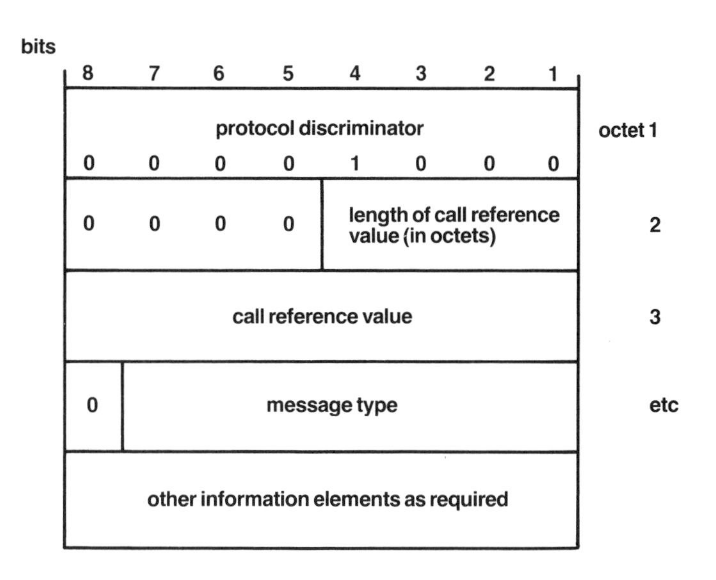

Figure 5.13
Signalling message structure.

work defines the use of ISDN for packet-switched calls in the D channel, for exchange of information between users, and the invoking of supplementary services. The general structure of Layer 3 signalling messages is shown in Figure 5.13.

The first octet contains a protocol discriminator which gives the D channel the capability of simultaneously supporting additional communications protocols in the future. The bits shown are the standard for user–network call control messages.

The call reference value in the third octet is used to identify the call with which a particular message is to be associated. Thus a call can be identified independently of the communications channel on which it is supported. This feature is particularly important in connection with incoming call offering procedures on a passive bus arrangement since the channel is only allocated to the called terminal after answer.

The message type code in the fourth octet describes the intention of the message (e.g. a SETUP message to request call establishment). These are listed in Table 5.1. A number of other information elements may be included following the message type code in the fourth octet. The exact contents of a message are dependent on the message type, however, the coding rules are open ended and in principle it is a simple matter to include additional information elements to satisfy any requirement which may be indentified in the future. By way of example, Table 5.2 lists the information elements which may be contained in a SETUP message. This is probably the most complex message yet defined.

Table 5.1
Message types.

8	7	6	5	4	3	2	1	
0	0	0	0	0	0	0	0	escape to nationally specific message type:
0	0	0	–	–	–	–	–	*Called established message*:
			0	0	0	0	1	— ALERTING
			0	0	0	1	0	— CALL PROCEEDING
			0	0	1	1	1	— CONNECT
			0	1	1	1	1	— CONNECT ACKNOWLEDGE
			0	0	0	1	1	— PROGRESS
			0	0	1	0	1	— SETUP
			0	1	1	0	1	— SETUP ACKNOWLEDGE
0	0	1	–	–	–	–	–	*Call information phase message*:
			0	0	1	1	0	— RESUME
			0	1	1	1	0	— RESUME ACKNOWLEDGE
			0	0	0	1	0	— RESUME REJECT
			0	0	1	0	1	— SUSPEND
			0	1	1	0	1	— SUSPEND ACKNOWLEDGE
			0	0	0	0	1	— SUSPEND REJECT
			0	0	0	0	0	— USER INFORMATION
0	1	0	–	–	–	–	–	*Call clearing messages*:
			0	0	1	0	1	— DISCONNECT
			0	1	1	0	1	— RELEASE
			1	1	0	1	0	— RELEASE COMPLETE
			0	0	1	1	0	— RESTART
			0	1	1	1	0	— RESTART ACKNOWLEDGE
0	1	1	–	–	–	–	–	*Miscellaneous messages*:
			0	0	0	0	0	— SEGMENT
			1	1	0	0	1	— CONGESTION CONTROL
			1	1	0	1	1	— INFORMATION
			0	0	0	1	0	— FACILITY
			0	1	1	1	0	— NOTIFY
			1	1	1	0	1	— STATUS
			1	0	1	0	1	— STATUS ENQUIRY

The message sequence for call establishment is shown in Figure 5.14. In order to make an outgoing call request, a user must send all of the necessary call information (i.e. called party number and supplementary service requests) to the network. Furthermore the user must specify the particular bearer service required for the call (i.e. speech, 64 kbit/s unrestricted or 3.1 kHz audio) and any terminal compatibility information which must be checked at the destination. A low layer compatibility information element may be used to specify low layer terminal characteristics such as

Table 5.2
SETUP message.

	Direction	Length (octets)	Comments
Protocol discriminator	Both	1	
Call reference	Both	2–3	
Message type	Both	1	
Sending complete	Both	1	Optional; included if user or network indicates that all information is included in this SETUP message
Bearer capability	Both	6–8	Indicates CCITT telecommunications service
Channel identification	Both	2–?	To identify a channel within an ISDN interface controlled by these procedures
Network specific facilities	Both	2–?	Optional
Display	Network–User	2–82	Optional: IA5 (ASCII) characters for display at terminal
Keypad	User–Network	2–34	Alternative for transferring called party number. Keypad may be used for other information
Calling party number	Both	2–?	Optional
Calling party sub-address	Both	2–23	Optional
Called party number	Both	2–?	In user–network direction alternative to use of keypad
Called party sub-address	Both	2–23	Optional
Transit network selection	User–Network	2–?	Optional
Low layer compatibility	Both	2–16	Optional
High layer compatibility	Both	2–4	Optional
User–user	Both	2–131	Optional, when the calling user wants to pass user information to the called user

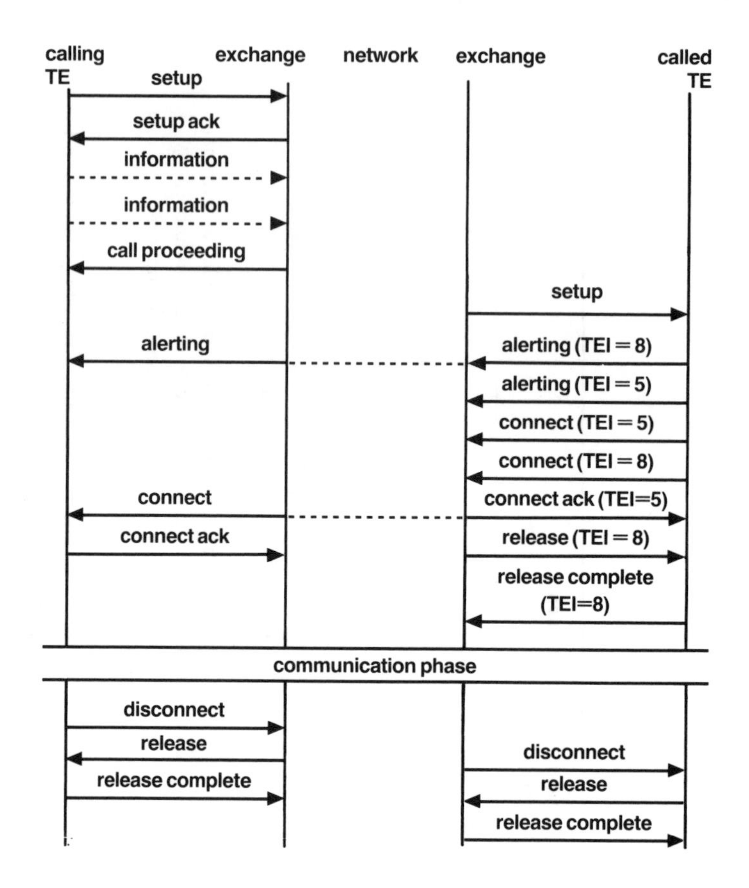

Figure 5.14
Message sequences for basic call establishment and clearing.

data rate. Where applicable the non-voice application to be used on the call may be specified via the high layer compatibility information element (i.e. Group 4 FAX, Teletex, Videotex or Slow Scan Video).

The initial outgoing call request may be made in an *en bloc* or overlap manner. Figure 5.14 illustrates the call establishment procedures. If overlap sending is used then the SETUP message must contain the bearer service request but the facility requests and called party number information may be segmented and conveyed in a sequence of INFORMATION messages as shown. Furthermore, if a speech bearer service is requested and no call information is contained in the SETUP message, then the network will return in-band dial tone to the user until the first INFORMATION message has been received.

Following the receipt of sufficient information for call establishment, the network returns a CALL PROCEEDING message to indicate that it is attempting to establish a connection to the called party.

At the called side a SETUP message is delivered to the called party via the broadcast data link. All terminals connected to the NT1 can examine the SETUP message to determine whether or not they are compatible with the calling party. This compatibility check must be performed by examining the bearer capability and low layer compatibility information element. Furthermore an indication of the application required (e.g. Group 4 Fax and Teletex) will be found in the high layer compatibility information element if this has been provided by the calling party. In this particular example two terminals have identified that they are compatible (TEI5 and 8) and return an ALERTING message to the network. At the same time, if appropriate, the terminal should generate a local indication of the incoming call (e.g. ring bell). When a terminal detects that the call has been answered it forwards a CONNECT message to the network. The network allocates the call to the first terminal to return a CONNECT and instructs that terminal to connect to the appropriate B channel by returning a CONNECT ACKnowledge containing the B channel identification. All other terminals which had responded to the incoming call will receive a RELEASE message from the network indicating that they should clear and return to the idle state. Following the receipt of a CONNECT message from a terminal at the called user's installation, the network will advise the calling party that the call has been answered by sending a CONNECT message. Call charging can then commence if appropriate.

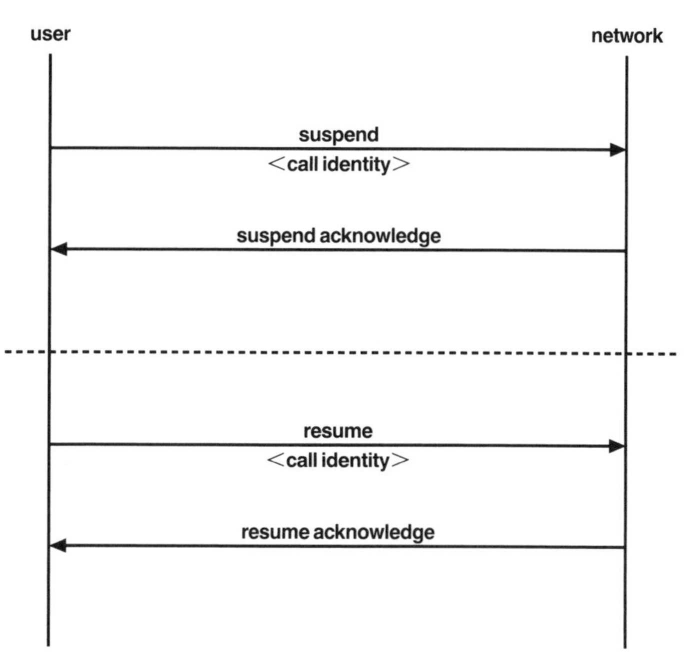

Figure 5.15
Suspend/resume procedures.

A user may clear a call at any time by initiating a three-message sequence (DISConnect, RELease and RELease COMPlete). In certain circumstances clearing may be initiated through the sending of RELease or RELease COMPlete.

In addition to call establishment and clearing it is possible to suspend a call in order to move a terminal between sockets on the passive bus. Following such an operation, the call may be resumed from the new position by sending a RESUME message to the network with the call identity in the form of two IA5 characters as defined by the user at the time of call suspension. The message sequences for suspend/resume are shown in Figure 5.15.

5.4 ENHANCEMENT OF BASIC CALL CONTROL PROCEDURES

David Davies

Two areas where there has been significant development of the call control protocol are the detailed procedures for terminal selection and in the procedures to support 7 kHz (enhanced quality) telephony and videotelephony services. Chapter 8 considers the implementation of such services. Here we consider the signalling needs.

5.4.1 Terminal selection

When setting up a call the customer may well have the choice of a multiplicity of terminals connected to the destination ISDN access and selection of the appropriate terminal is of some importance. Firstly there may be several terminals of the same type (e.g. telephones) all connected to the same access and the caller may wish to select a particular one, on the basis of his knowledge of its location. Secondly it is important to ensure that only a compatible terminal is connected. Thus if the originating terminal is a facsimile terminal then a facsimile terminal must be selected at the destination. Situations will arise where both selections are relevant, for example where a basic rate interface has three telephones and three associated facsimile machines connected. In this case the caller will have an opportunity to select the particular terminal from the three and the signalling will ensure that the appropriate type of terminal is connected. Let us now consider in more detail the two processes whereby these selections are made.

Addressing a specific terminal

Two mechanisms exist for addressing a specific terminal. The first uses the supplementary service Direct Dialling-In (DDI) which is called Multiple Subscriber Number (MSN) when applied to ISDN basic access. In both DDI and MSN, part of the public network number is used for routing the call within the customer's premises. In the case of a passive bus configuration on an ISDN basic access, the MSN supplementary service will provide the relevant part of the public network number (i.e. the last digit(s) which have not been used for routing the call through the public network) to be sent in the incoming SETUP message which is broadcast to all terminals on the passive bus. Each terminal on the passive bus must be allocated a different MSN number which is programmed into the call handling process in the terminal. Only that terminal with an MSN number matching the MSN number included in the incoming SETUP message will accept the call, provided that its terminal characteristics are compatible with those specified in the compatibility information elements contained in the SETUP message and that it is in a state to accept the call. This is illustrated in Figure 5.16.

The second mechanism for addressing a particular terminal uses the SUB-addressing (SUB) supplementary service. In this supplementary service, additional addressing information (the sub-address) is conveyed transparently from calling to called terminal. This address is not part of the ISDN number used by the network for routing the call. In full OSI (see Section 1.10) applications, it can be used to address a specific process within the terminal but it can also be used in non-OSI applications. Each terminal connected to the passive bus needs to be allocated a sub-address

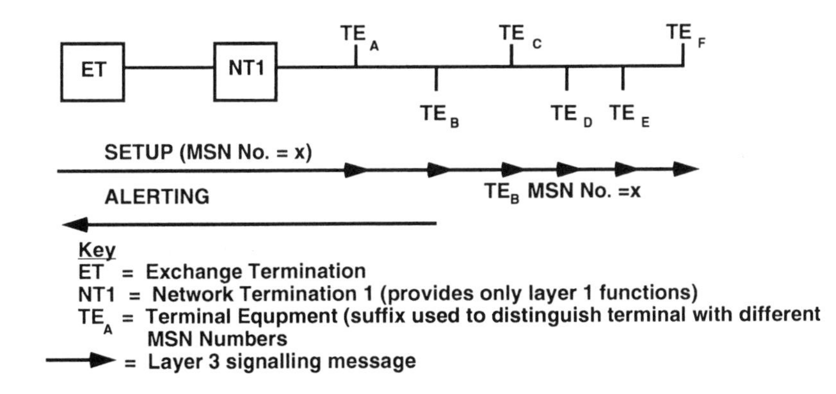

Figure 5.16
Terminal selection on a passive bus using MSN.

which is programmed into its call handling process. The incoming SETUP message which is broadcast to all terminals on the passive bus will contain the called sub-address and only the terminal with the matching sub-address will respond to the incoming call.

The significant difference between DDI/MSN and SUB supplementary services is that for DDI/MSN the address allocated to the terminal is part of the public ISDN number whereas for SUB the address is not part of the public ISDN number. Hence the use of DDI/MSN needs to be carefully controlled by the network operator to ensure that the public ISDN number space does not become exhausted. In addition, DDI/MSN is the only mechanism that can be used for selecting a specific terminal in the case of interworking with the PSTN (public switched telephone network) since the network signalling system of the PSTN does not allow the conveyance of any addressing information other than the public number, i.e. it cannot convey sub-addressing information.

Terminal compatibility

The access and network ISDN signalling systems allow the calling user to indicate, in the originating SETUP message, telecommunications requirements of the connection to be provided by the network using the Bearer Capability (BC) information element and the characteristics of the terminal using the High Layer Compatibility (HLC) and Low Layer Compatibility (LLC) information elements. At the called side, these information elements are sent in the incoming SETUP message (when provided by the calling terminal) and are used by the called terminal to determine if it is compatible with the calling terminal and hence whether to accept or reject the call. This compatibility checking mechanism may be used in conjunction with the addressing mechanisms discussed above.

5.4.2 Bearer service

The BC is used to indicate to the network the characteristic of the connection (bearer) required by the calling terminal or, in the case of the incoming SETUP message at the called side, the characteristics of the connection allocated by the network for the call. Typical examples are:

speech,
3.1 kHz audio,
64 kbit/s unrestricted.

Note that 'BC = speech' would indicate to the network that the normal routing rules for speech calls apply (i.e. no more than two satellite links, echo control devices may be required, Digital Circuit Multiplication Equipment (DCME) may be used to only allocate an expensive intercontinental channel when speech is actually being spoken, A/μ-law conversion required on connections between North America and elsewhere) and that interworking with the PSTN is allowed.

'BC = 3.1 kHz audio' is used for modem applications so in this case echo control devices and DCME should not be used (note that on the PSTN, modem disable these devices by the application of 2100 Hz tone which is applied as soon as the connection is established). Again A/μ-law conversion is required when operating between countries using μ-law and A-law encoding and interworking with the PSTN is allowed.

'BC = 64 kbit/s unrestricted' requires the network to provide an all digital, 64 kbit/s, bit-transparent path. Call progress tones and announcements will not be provided in this case (call progress information is provided via the D-channel signalling) and interworking with PSTN is precluded.

If provided by the calling user, the HLC and LLC information elements are transported transparently across the network. The HLC provides information about the higher layer functionality of the terminal (e.g. telephony, Group 3 FAX, Group 4 FAX, Teletex, etc) whilst the LLC provides information about the lower layer characteristics of the terminal (e.g. user rate, rate adaptation mechanism used, asynchronous or synchronous operation, parity information, etc).

In the case of speech or 3.1 kHz audio bearer service, interworking with the PSTN is allowed. For calls originating in the PSTN and destined for an ISDN terminal, the ISDN gateway exchange has to generate a BC since none of the compatibility checking information (i.e. BC, HLC or LLC) can be conveyed by the PSTN signalling systems. Hence at the PSTN/ISDN gateway, a BC = 3.1 kHz audio will be generated together with a progress indicator specifying that PSTN interworking has occurred and this information is conveyed to the called ISDN terminal. The progress indicator is used by the called terminal to modify its compatibility checking process so that no HLC or LLC is expected and that a BC = 3.1 kHz audio is considered to be compatible even if a BC = speech is expected.

All three information elements (BC, HLC and LLC) have many information fields associated with them in order to accommodate the large variety of requirements needed for the many types of terminals that can be supported on the ISDN. The international Standards specify how each of these fields will be coded for the various types of terminals and this has been necessary to ensure the operation of the international ISDN service. Such a specification can only cover the most common types of terminals and it does not preclude the use of other codings for terminals with characteristics not covered in the standard.

5.4.3 7 kHz telephony

7 kHz telephony makes use of the 64 kbit/s unrestricted digital connection which can be provided by the ISDN together with improved speech encoding techniques to provide a high quality speech service. The encoding process is discussed in Chapter 8.

The aspect of the 7 kHz telephony service which has demanded ingenuity in the standardization forum is the requirement that, if requested by the calling user, 'fallback' to operate at the normal 3.1 kHz bandwidth is allowed. The problem does not concern the inband encoding since the 7 kHz telephony encoding mechanism (defined in CCITT Recommendations G.722 and G.725) has been designed to have an alignment sequence at the start of the connection and if alignment is not achieved then A- or μ-law encoding (whichever is appropriate to the network to which the terminal is attached) is assumed. The problem relates to how to enhance the signalling system to allow negotiation of the 7 kHz/3.1 kHz services which will not require any change to speech (i.e. 3.1 kHz) terminals designed to the simple standard. In addition, there are different network requirements depending on whether the resultant call ends up as a 7 kHz speech call, since for 7 kHz telephony echo cancellation will be carried out in the terminal and the network must not include in the connection any DCME or echo control devices (since they are designed to operate on the 3.1 kHz A/μ-law encoding scheme). The requirements established for the design of the negotiation or 'fallback' signalling procedure are:

- it must not require any changes to existing speech (3.1 kHz) terminals or any changes to the network procedures for handling 3.1 kHz speech telephony;
- it must be sufficiently fast to give an acceptable call set-up time;
- it must be generic (i.e. can be used for services other than 7 kHz telephony);
- it must signal to the network and calling user the result of the negotiation, i.e. whether 'fallback' has occurred.

The international Standards specify the following procedures for handling fallback:

- when no 'fallback' or interworking with PSTN is allowed, the calling 7 kHz telephony terminal will set BC = 7 kHz (G.722/G.725) in the outgoing SETUP message;
- when 'fallback' and interworking with PSTN is allowed, the calling 7 kHz telephony terminal will include two BC information elements in

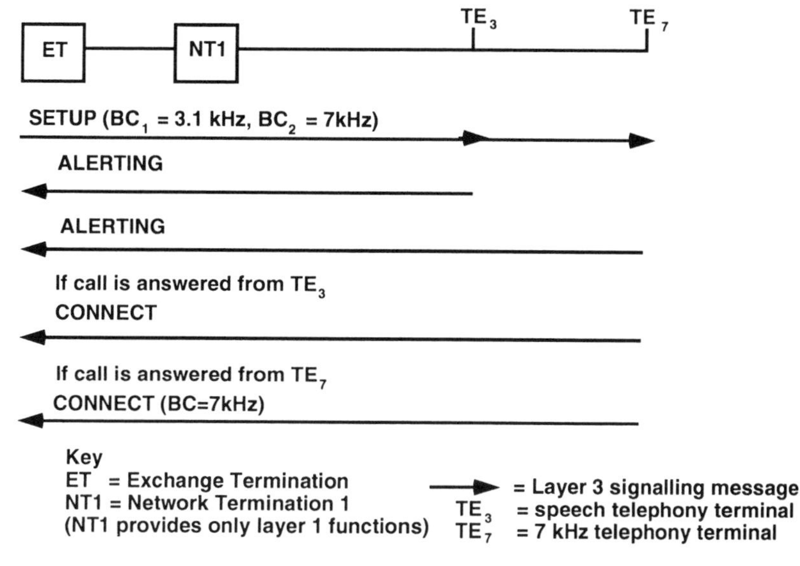

Figure 5.17
Call offering in the case of 7 kHz telephony with 'fallback' allowed.

the outgoing SETUP message; the order of the BCs will be in ascending order of priority. Hence the first BC_1 = speech (G.711) and the second BC_2 = 7 kHz (G.722/G.725). The procedure at the called side is illustrated in Fig. 5.17.

At the called side, the SETUP message containing the two BCs is broadcast to all terminals on the passive bus. Existing speech (3.1 kHz) terminals which only expect to receive a single BC in the incoming SETUP message, will ignore the second BC and process only the first BC. Hence, provided the remaining contents of the SETUP message (e.g. MSN number) is compatible with the terminal, the 3.1 kHz telephony terminal will respond to the incoming call. 7 kHz telephony terminals will be implemented with the new version of the protocol and will thus process both BCs and respond accordingly. A further enhancement to the protocol is that when an incoming SETUP message contains more than one BC, a terminal answering a call will include in the CONNECT message the BC accepted by the terminal unless it is the first BC in the SETUP message. Hence for the 7 kHz telephony terminal, the terminal will include BC = 7 kHz (G.722/G.725) in the CONNECT message. If no BC is included in the CONNECT message, the network will assume that 3.1 kHz telephony is required. In either case, the CONNECT message to the calling terminal will contain the BC of the resultant connection. If both terminals are answered simultaneously, the network will allocate the call to the terminal from which the CONNECT message is received first.

5.4.4 Videotelephony

There are two cases identified for the videotelephony service as a result of the videotelephony encoding mechanism discussed in Chapter 8.

(1) videotelephony based on using one B-channel at the user–network interface,

(2) videotelephony based on using two B channels at the user–network interface. There are two variants of this case; one where the audio signal is sent over one B channel and the video signal over the other and the other variant is where the audio signal occupies only part of one B channel and the video signal occupies the remainder of that B channel and the whole of the second B channel.

The enhancement which are required to the signalling protocols to support the videotelephony service are the provision of:

- additional codings in BC and HLC for videotelephony,
- signalling procedures to allow 'fallback' to 3.1 kHz or 7 kHz telephony,
- for case 2, a mechanism to establish 2 B-channel connections to the same user (ie 2 × 64 kbit/s bearer service).

The first enhancement is easy to resolve and the second requirement is very similar to the 7 kHz to 3.1 kHz telephony 'fallback' procedure discussed in Section 5.4.3 and is accommodated using the same signalling procedure.

The provision of a 2 × 64 kbit/s connection is more difficult since it is a requirement that each 64 kbit/s connection has similar characteristics particularly in terms of delay, i.e. it is not acceptable for one connection to be routed via a satellite and the other by a terrestrial link. The terminals themselves have to ensure bit sequence integrity over the 2 × 64 kbit/s connection as this cannot be provided on existing 64 kbit/s switched networks. One possibility is that the terminal establishes both B channel connections as two separate calls but the current access signalling Standards does not allow the request of quality of service parameters to ensure that a similar routing is achieved for both connections. The best solution would be to enhance the signalling protocol to allow the request for both B channels to be established to the same connection using a single SETUP message but, whilst this may be a fairly easy enhancement for the access signalling system, it would still require enhancements to the network inter-exchange signalling system and to the exchange call control procedures which currently establish 1 × 64 kbit/s connections independently of each other. Further discussion of the problem of synchronizing many channels is given in Chapter 9.

5.4.5 Access protocols for supplementary services

In order to make ISDN an attractive service offering to users, it must support many supplementary services, particularly those that are typically available on PABXs and private networks. Hence many man years of effort have been expended to specify all the Standards that are necessary to implement ISDN supplementary services on an international basis.

For the user–network interface, two types of generic Layer 3 signalling protocols have been specified for the control of supplementary services, namely functional and stimulus signalling procedures.

In stimulus signalling procedures, the terminal is not required to have any knowledge of the supplementary service being invoked. No records of the supplementary service call state are held by the terminal and Layer 3 signalling messages are generated as a result of human action (e.g. pressing a button on the keypad). The operation of the terminal (e.g. the display of messages, lighting of indicator lamps) is controlled by the network via Layer 3 signalling messages. Hence the network and terminal operate in a master/slave relationship and no intelligence is required in the terminal.

In functional signalling, the terminal must have knowledge of the supplementary service being invoked and have the associated signalling protocol implemented. Both the terminal and network hold records of the supplementary service call state.

Stimulus signalling

There are two types of stimulus procedures defined—the keypad protocol and the feature key management. Both stimulus procedures use the basic call control Layer 3 messages, particularly the INFORMATION message, as the transport mechanism for conveying the stimulus information between the terminal and the network. In the keypad protocol, the user invokes a supplementary service by keying in the appropriate sequence of digits delimited by * and # in the same manner as services on analogue telephones using multifrequency keypads (see Figure 5.18). The generic procedure has been specified and the sequences of *, # and digits for any given supplementary service is network dependent.

The feature key management protocol requires the network to hold a terminal or service profile for a given terminal. A given supplementary service is allocated a feature number and is invoked by the terminal

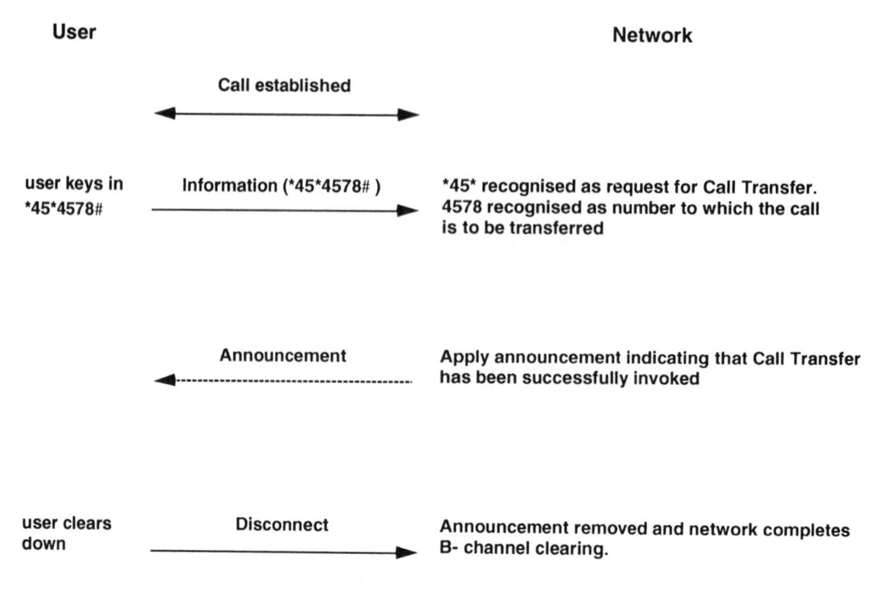

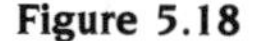

Figure 5.18
Example of operation of keypad protocol for invoking supplementary services.

signalling that the feature number to the network, e.g. a pre-programmed key on the terminal marked 'call transfer' will generate an INFORMATION message containing feature activation number 3 (for example) which corresponds to the call transfer supplementary service in the service profile held for that terminal in the local exchange (see Figure 5.19). This mechanism is very similar to that currently offered on PABXs where the PABX terminals have feature keys which can be programmed to correspond to the supplementary service required by the user.

Both stimulus procedures can easily handle many of the supplementary services involving only a single call but are unsuitable for those supplementary services involving more than one call, e.g. conference calls. In addition, it is desirable that for PABXs, the same generic protocol could be used whether the PABX was operating to a public network or direct to another PABX in a private network configuration. In order to meet these requirements, it was necessary to define a functional protocol and it is the functional protocol which has been specified not only as a generic protocol but also for each ISDN supplementary service defined by CCITT. Hence it is intended that the functional protocol is specified in detail for all defined ISDN supplementary services and the stimulus protocol will be used either for an early introduction of these supplementary services or for the control of network- or manufacturer-specified supplementary services (i.e. non-CCITT defined services).

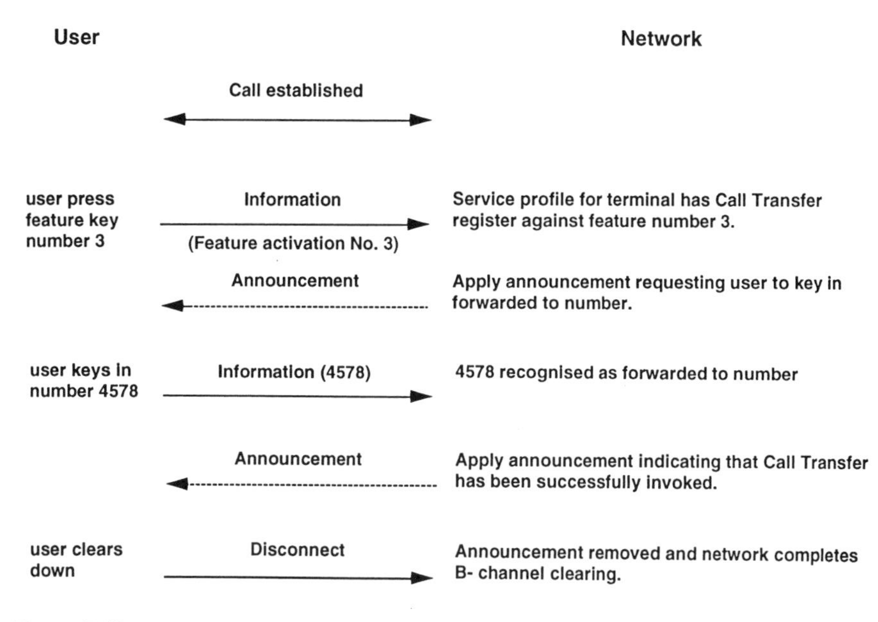

Figure 5.19
Example of operation of Fecture key management protocol for invoking supplementary services.

Functional signalling

The functional protocol is based on using the facility information element which is conveyed in the basic call control messages if the supplementary service is invoked during either call establishment or call clearing and in the FACILITY message otherwise. For the control of supplementary services which are independent of an active call, the REGISTER message is used to convey the facility information element. In addition, specific Layer 3 messages have also been defined for the function of holding and retrieving a call.

The protocol applicable to the supplementary service information contained in the facility information element is based on the ROSE (Remote Operations Service Element) protocol. The ROSE protocol is used to support applications where interactive control of objects is required in an open systems environment.

The ROSE protocol is used in the context of the ISDN functional supplementary services to control functional entities (application processes) within network (either local or remote to the user). The ROSE protocol was seen as appropriate for use in the context of the supplementary services since the following operations are required in the control of supplementary services and these are provided by the ROSE protocol:

- initiation of the service process,
- confirmation response (either positive or negative),
- further control after initiation.

The ROSE procedure specified by CCITT is self-contained in that it contains two methods of transportation of the ROSE APDUs (Application Protocol Data Unit). However, for the purpose of supplementary service control over the user–network interface, the ROSE APDUs are transported within the facility information element within one of the Layer 3 signalling messages as described above. Hence some simplification of the full ROSE protocol has been possible.

The supplementary services make use of the ROSE procedures in the following way:

(a) An entity (usually the user) initiates a supplementary service by sending an RO-INVOKE (ROIV) APDU, e.g. by pushing a particular button which could be labelled and would generate a predefined message to the network in a Q.932 facility information element. This initiates the appropriate state machine for that supplementary service.

(b) The remote entity (the supplementary service state machine usually located in the network) tells the initiator to continue or abort by sending an RO-RESULT (RORS) APDU (to continue), an RO-ERROR (ROER) APDU (to say that the invocation was syntactically incorrect and abort) or an RO-REJECT (RORJ) APDU (to reject the invocation and abort).

(c) Further control of the service by the initiator is performed by further ROIV APDUs as appropriate for the supplementary service.

An example of the application of the ROSE protocol to the ISDN is the Call Hold (HOLD) supplementary service. When there is an active call at an access and the user wishes to put a call on hold to do something else, then a HOLD message is sent to the remote entity (the network). This HOLD message is the initiation of the service process and is equivalent to the RO-INVOKE APDU. The remote entity tries to perform the operation requested and sends an appropriate response—either a HOLD ACKNOWLEDGE message in the case that the operation was successful (equivalent to the RO-RESULT APDU), or a HOLD REJECT message in the case that the operation was unsuccessful (equivalent to the RO-REJECT APDU) (see Figure 5.20). A similar procedure applies to retrieval of the call with the RETRIEVE, RETRIEVE ACKNOWLEDGE and RETRIEVE REJECT messages. This example shows how the *principle* of the ROSE protocol has been used for a particular application.

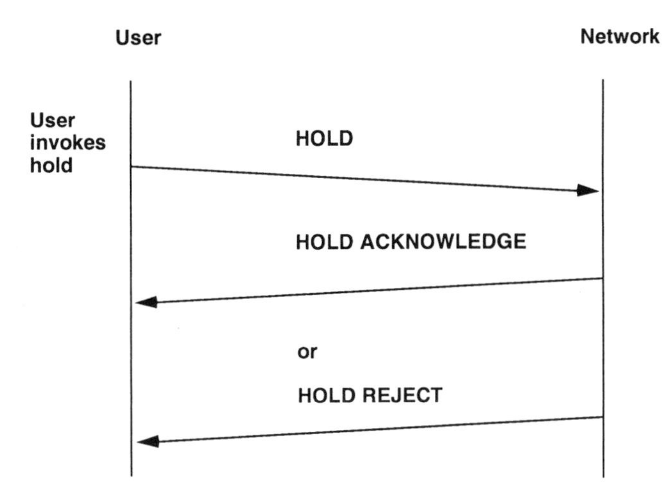

Figure 5.20
The HOLD supplementary service.

Another example is the Malicious Call Identification (MCID) supplementary service. A user wishing the network to register a malicious call, sends a Facility Information Element (IE) to the remote entity (the network). This information element contains an 'MCID Invoke' component (equivalent to the RO-INVOKE APDU). The remote entity will try to perform the opera-

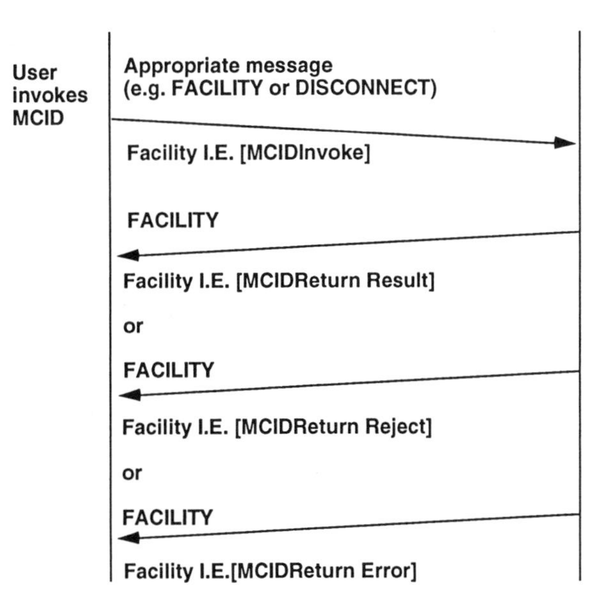

Figure 5.21
The MCID supplementary service.

Table 5.3
Supplementary services.

Calling line identification presentation	User–user signalling
Calling line identification restriction	Advice of charge
Connected line identification presentation	Completion of calls to busy subscribers
Connected line identification restriction	Conference call, add-on
Closed users group	Meet-me conference
Call waiting	Call forwarding unconditional
Direct dialling in	Call forwarding busy
Malicious call identification	Call forwarding no reply
Multiple subscriber number	Call deflection
Sub-addressing	Freephone
Terminal portability	Three-party service

tion and, dependent on the outcome, will respond accordingly. If the operation was successful then the remote entity will respond with a 'MCID Return Result' component (equivalent to the RO-RESULT APDU), if the operation was unsuccessful or could not be performed then the remote entity will respond with an 'MCID Return Reject' component (equivalent to the RO-REJECT APDU) and if the Invoke component was incorrectly structured then the remote entity will respond with an 'MCID Return Error' component (equivalent to the RO-ERROR APDU) (see Figure 5.21). This example shows how the ROSE protocol has been used in the supplementary services.

Other supplementary services make use of the ROSE protocol in similar ways, some of them just using the *principle* and others using the components as described above.

ISDN supplementary services specified using the functional protocol are listed in Table 5.3.

5.5 RELATIONSHIP WITH CCITT SIGNALLING SYSTEM No 7

In Section 3.5 the use of CCITT No 7 in the IDN was discussed. The Layer 3 ISDN signalling between the terminal and the local exchange is carried across the network by CCITT No 7 and at the distant end is reconverted to the ISDN signalling. Figure 5.22 shows the message flows of a possible ISDN call set-up. Note that the diagram shows the case where the call set-up information is sent from the customer in an overlap fashion, rather than *en bloc* in a single SETUP message. Transit exchange 2 has sufficient information that it knows how much routing information is needed and hence reassembles it into a single IAM. Some information is only transfer-

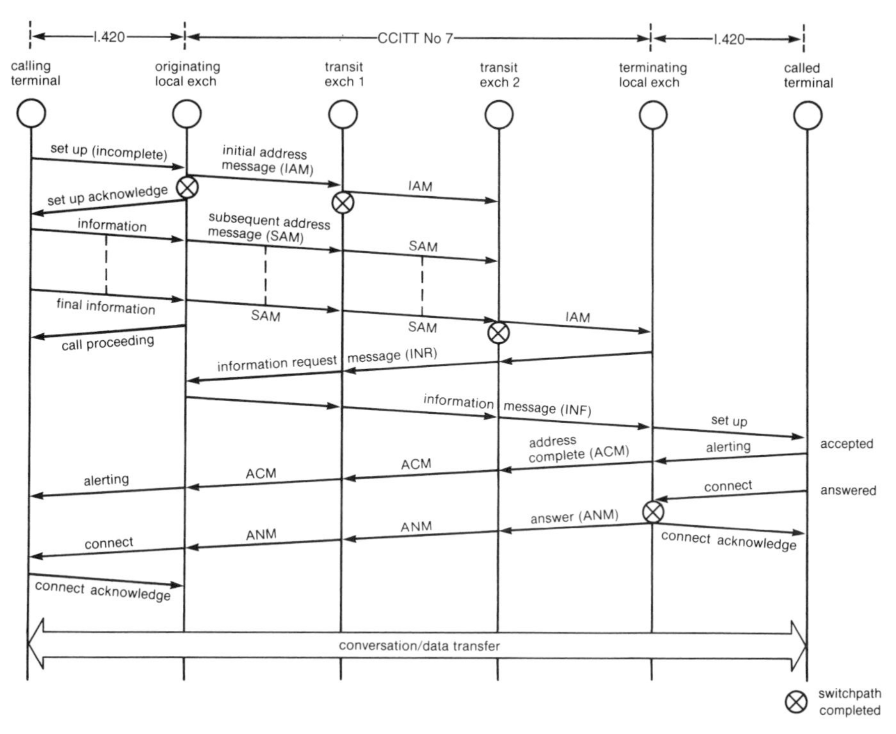

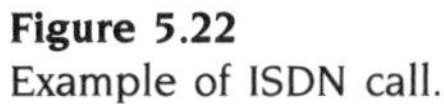

Figure 5.22
Example of ISDN call.

red to the terminating exchange when it is ready to receive it, requested by the INR message and returned in the INF message.

5.6 MAINTENANCE

It is essential that the network administration has some means whereby it can centrally test customers' connections. On analogue links to the customer d.c. or low frequency tests are customary. Historically this has been provided over metallic copper test pairs from an exchange test desk. Even for analogue links this is becoming less desirable as it inhibits the use of remote subscriber interface units operating over multiplexed links particularly where the physical connection is provided by optical fibres. When the final link to the customer is digital, the use of metallic pairs to provide a test link is even less satisfactory as digital streams observed over a test link of any length are so modified by the attenuation of the link that useful information is not obtained. For this reason testing of the ISDN is based on the principle of looping back signals from the exchange in the terminal

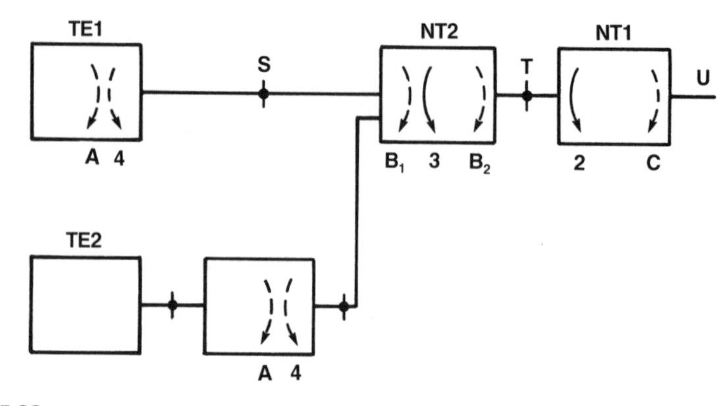

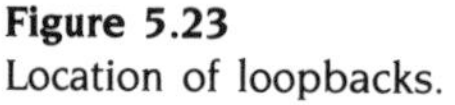

Figure 5.23
Location of loopbacks.

apparatus so that the signals are returned to the exchange. Other loopbacks could return the channel to the user. There they can be observed and any errors monitored.

Figure 5.23 shows possible locations of loopbacks. The dotted loopbacks are regarded as optional but loopbacks labelled 2 and 3 are 'recommended' and 'desirable' respectively. Loopback 2 is in the NT1 as near as possible to the T reference point. It will cause B channels to be looped back and the D channels also, in so far as that is practical. It will be under the control of Layer 1 signal from the exchange. Loopback 3 is as near as possible to the S reference point and may either be used for local maintenance or be controlled by Layer 3 messages in the D channel.

5.7 TESTING AND APPROVALS

As may be imagined, proving that equipment performs correctly to the protocols described in this chapter is important. There are several needs.

(a) The initial specification must be complete, self-consistent and unambiguous. This has always been a serious problem and the tendency is to go to more formal methods to describe the actions of terminal and network, with a mathematically rigorous background. Hence much of the protocol description in the CCITT documentation is in the graphical Specification and Description Language (SDL) as described in Appendix A. More formal is the ISO language LOTOS.

(b) The equipment must be tested for conformance to the specification. The protocols are so complex that it is theoretically impossible to investigate every possible sequence of messages, but as a minimum every possible path through a call must be explored. However, this is not sufficient; it is important that non-permitted sequences be applied to equipment to ensure that no mechanism is available which would cause systems to lock-up or in some other way degrade the network. The non-permitted sequences must be ignored or rejected to protect against data corruption on the line or faulty apparatus. Because of the large number of sequences involved, a powerful processor with software to match and the necessary hardware interfaces is needed. For Layer 1 testing an automated rig of oscilloscopes, power supplies, spectrum analysers and network analysers is needed.

(c) The equipment performance must be checked. It is not sufficient for equipment simply to conform to specification at low calling rates, but this conformance must be maintained at the design call rate and under overload. It is only too easy for buffers to overflow into one another in RAM causing unpredictable results which will not show up in conformance tests. Test gear must be able to operate at such speeds as will show up these effects.

(d) The approvals regime must be established to ensure that only satisfactory equipment is connected to the network. It may be felt that full conformance is not necessary but as a bearest minimum electrical safety is necessary. Approval to connect to a public ISDN network interface must be obtained by a manufacturer.

In Europe the approval specifications are known as 'Normes Européennes de Télécommunication' (NET). The following NETs are relevant.

NET3 — Basic rate ISDN connection.
NET5 — Primary rate ISDN connection.
NET7 — Terminal adapters.
NET33 — Digital telephony.

The guiding principle under which the NETs are written is the need to ensure that essential requirements are met. These include user safety, the safety of employees of public telephone network operators, the protection of networks from harm and (in justified cases) the interworking of terminal equipment.

In the USA approval must be sought under the Federal Communications Commission (FCC) Rules Part 68, the objectives of which are very similar to the NETs. Part 15 covers radio-frequency radiation.

REFERENCES

ISO Standard 8877. Interface connector and contact assignments for ISDN basic access at reference points S and T.

CCITT Recommendations:

G.722	7 kHz audio coding within 64 kbit/s.
G.725	System aspects for the use of 7 kHz audio codec within 64 kbit/s.
I.430	Basic user–network interface—Layer 1 specification.
Q.920	ISDN User–network interface data link layer general aspects.
Q.921	ISDN User–network interface—data link layer specification.
Q.930	ISDN User–network interface Layer 3—general aspects.
Q.931	ISDN User–network interface Layer 3 specification for basic call control.
Q.932	Generic procedures for the control of ISDN supplementary procedures.
X.219	Remote operations: model, notation and service definition.
X.229	Remote operations: protocol specifications.
Z.100–104	S.D.L.

Other documents are supplied by network operators which will include details of national variatiants and options. Examples are:

In the UK—British Telecom Network Requirement (BTNR) 191. I series interfaces for ISDN access.
In Germany—FTZ – Richtlinie 1R6.
In the USA—ANSI Standards:

T1.604	Minimal set of bearer services for the basic rate interface.
T1.602	Basic rate interface, Primary Rate interface link access procedure, D-channel (LAP1): BRI/PRI Data Link Protocol.
T1.605	Basic rate interface specification at the user–network interface: BRA S/T interface specification.
T1.607	Basic call control procedures—BRI and PRI.
T1.608	Supplementary Services Control Procedures—BRI and PRI.
T1.609	Interworking between the ISDN user–network interface protocol and Signalling System No 7 ISDN user part.
T1.610	Generic procedures for the control of ISDN supplementary services.

In France—France Telecom Spec VN2.

NETs are published by CEPT Liaison Office, Seilerstrasse 22, CH-3008 Bern.

For more details, particularly of CCITT No 7 Signalling, see *Common Channel Signalling* by R. Manterfield, published by Peter Perigrinus.

QUESTIONS

1 How many pins are used on the ISDN connector? What are they used for?

2 The incoming information rate from the exchange is 144 kbit/s but the ISDN interface operates at 192 kbit/s. How are the extra bits used?

3 Starting from an idle circuit, how is a simple telephony call initiated including the operation of all layers?

4 What is the difference between stimulus and functional signalling and what are the advantages of each?

Chapter 6

Primary Rate ISDN Access

6.1 BACKGROUND

Domestic and small business customers having a need for a small number of connections at 64 kbit/s will be served with ISDN by the basic rate access described in Chapter 5. However, larger customers may require access to tens or hundreds of connections at 64 kbit/s for PABX or computer bureaux type service; as this implies both voice and data switching the more generic term Private Telecoms Network Exchange or PTNX can be used. For such an exchange some way of offering service in larger blocks is required. For this reason the interfaces used within the network have been modified and pressed into service. The interfaces operate at 1.544 Mbit/s in North America and 2.048 Mbit/s in Europe and the Layer 1 format is described in Section 3.2. In addition to the provision of multichannel ISDN access to customer, the primary rate interface may be used to access remote units providing basic rate access to customers (see Section 3.4).

The frame structures for ISDN access are shown in Figures 6.1 and 6.2. Note that the D signalling channel is at a rate of 64 kbit/s and is in timeslot 24 in the 1.544 Mbit/s system and timeslot in 16 in the 2.048 Mbit/s system. It is possible that one D channel may carry signalling for more than one primary rate interface in which case all 24 and 31 timeslots can be used for B channels. The electrical characteristics of the interfaces are pulses of 3 V nominal amplitude on a symmetric pair into a resistance of 120 Ω (2.048 Mbit/s) or 100 Ω (1.544 Mbit/s). An alternative interface in use at 2.048 Mbit/s in the United Kingdom is 2.37 V peak nominal on a coaxial line of 75 Ω impedance. The pulses are half-width; that is to say that each pulse only occupies half the pulse period (see Figure 3.3), unlike the pulses on the basic rate interface which are full-width. The coding is HDB3 (2.048 Mbit/s) and B8ZS (1.544 Mbit/s) as described in Section 3.2. It must be emphasized that these are the signals presented to the customer. The signals on the transmission line to the customer may be the same, or may be quite different. For example in the UK a more complex 4B3T line code may be used which, by reducing the modulation rate of a 2.048 Mbit/s signal to 1.536 Mbaud, can extend the reach available to the customer. Alternatively optical fibre transmission may be used which employs a totally different coding technique.

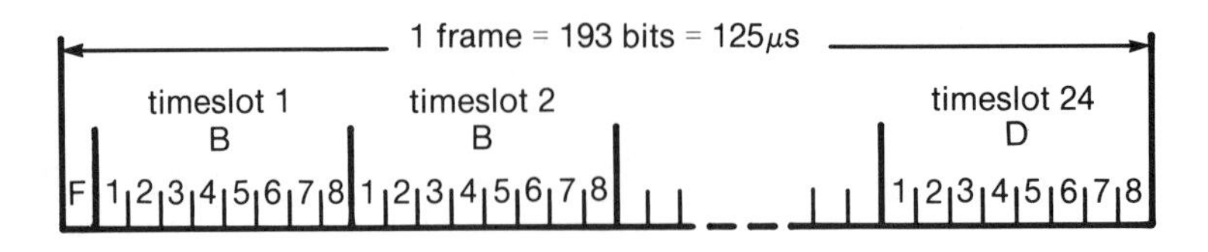

Figure 6.1
Frame structure of 1.544 Mbit/s primary rate interface.

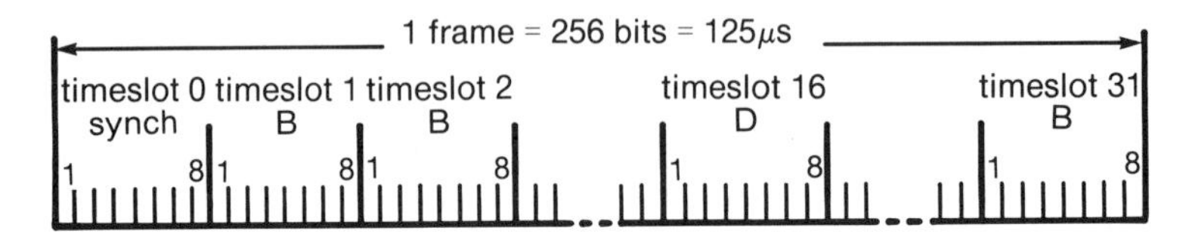

Figure 6.2
Frame structure of 2.048 Mbit/s primary rate interface.

6.2 SIGNALLING

Differences between the primary rate interface and the basic rate interface are:

(a) No provision is made in the primary rate interface for multipoint working in the customer's premises; the connection will simply be between the network and a single PABX or other piece of equipment. Of course the individual channels across the interface can be routed independently to whatever destination is required.

(b) No provision is made in the primary rate interface for deactivating the link to economize in power.

(c) Any customer who uses as many circuits as provided by the primary rate access may be expected to have both switched and private circuits and it is desirable that the interface be able to support a mix of both.

The first two points simplify design of the signalling protocols. It is relatively straightforward to adopt the basic rate protocols for use in primary rate service. The 64 kbit/s D channels provide additional throughput to support the larger number of channels. Such services were introduced in West Germany in 1988. In the USA AT&T introduced such a service for access to its long haul network services. Australia, Belgium, Finland, France and Japan were close behind, as were the Local Exchange Carriers in the USA. The final point on the mixing of private and public switched connections is more difficult.

6.3 EVOLUTION OF PABX SIGNALLING

In Section 1.8 mention was made of the pilot ISDN service offered by British Telecom in 1985. Its specification was based on discussions which were taking place at CCITT forums around 1982. Hence they incorporated ideas which were later developed to form the 1984 Red Book and 1988 Blue Book Recommendations. The signalling used on the pilot service basic rate access was called DASS (Digital Access Signalling System). This was subsequently developed in two ways.

(1) As DASS2, to provide for primary rate access to the ISDN. This was used as the signalling interface to the primary rate ISDN service in the UK until supplanted by I.421 in the early 1990s.

(2) As a Digital Private Network Signalling System (DPNSS) for the interworking of PABXs over private circuits introduced in 1985.

Owing to their common ancestry the two forms of signalling can coexist in channel 16 of the primary rate access and so private inter-PABX circuits can be mixed with channels connected to the public switched network on the same 30 channel multiplex.

The Layer 2 protocols are HDLC based. The frame is shown in Figure 6.3. The signalling operates in a 'compelled' mode. That is to say frames are transmitted continuously until an acknowledgement of correct reception is received from the distant end. Layer 2 frames containing messages on behalf of each of the 30 traffic channels are interleaved on channel 16. Access to channel 16 is given on a cyclic basis, thus giving equal priority to each traffic channel.

Layer 3 provides for call control. In principle the main difference between DASS2 and DPNSS lies in the symmetry of operation. DPNSS intercommunicates directly between PABXs of equal rank in the hierarchy, whereas DASS2 is used to communicate with a large national and international network.

DPNSS provides for interconnection to be made between PABXs of different manufacture, so that to the user the PABXs will appear as a single large PABX with all supplementary services available wherever the extension. DPNSS was the first 'open standard' digital inter-PABX signalling

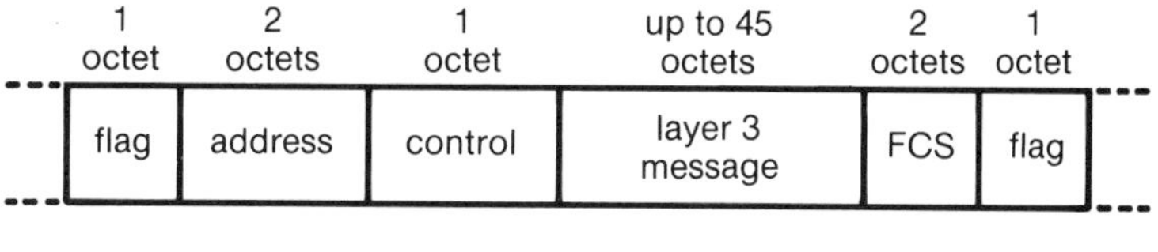

1 octet	2 octets	1 octet	up to 45 octets	2 octets	1 octet
flag	address	control	layer 3 message	FCS	flag

Figure 6.3
DASS2/DPNSS Layer 2 frame.

system, and has been implemented by many PABX manufacturers. It is therefore to be found far beyond the UK where it originated. However its lack of compatibility with the I.421 signalling Standard is inconvenient and there was a clear need for a successor. In response to this other proprietary Standards based on I.421 have also emerged such as TeLinc from Fujitsu and Telecom Australia, and Corenet from Siemens and Alcatel. However the problem is now being addressed internationally.

6.4 INTERNATIONAL STANDARDS FOR INTER-PABX SIGNALLING

ECMA, the European Computer Manufacturers Association, has taken the lead in this area with a signalling system called QSig. The title derives from the specification of a Q reference point within a PABX which terminates the inter-PABX network signalling. This is not expected to be a physical interface but is shown in conceptual terms in Figure 6.4. The physical interface is at point C, which could, for example, be a 1.5 or 2 Mbit/s digital interface. Another new piece of terminology is the 'Private Telecoms Network Exchange' or PTNX. This breaks away from the tradition that PABXs have of being primarily provided to offer voice services.

The QSig protocol essentially adopts the Layer 1 and 2 of I.421 but creates a Layer 3 more appropriate for inter-PTNX signalling, including the use of tandem switching. Figure 6.5 shows the setup of a simple call. This may be compared with Figure 5.22 which shows the equivalent call set-up process across the public network. The most noticeable difference is use of

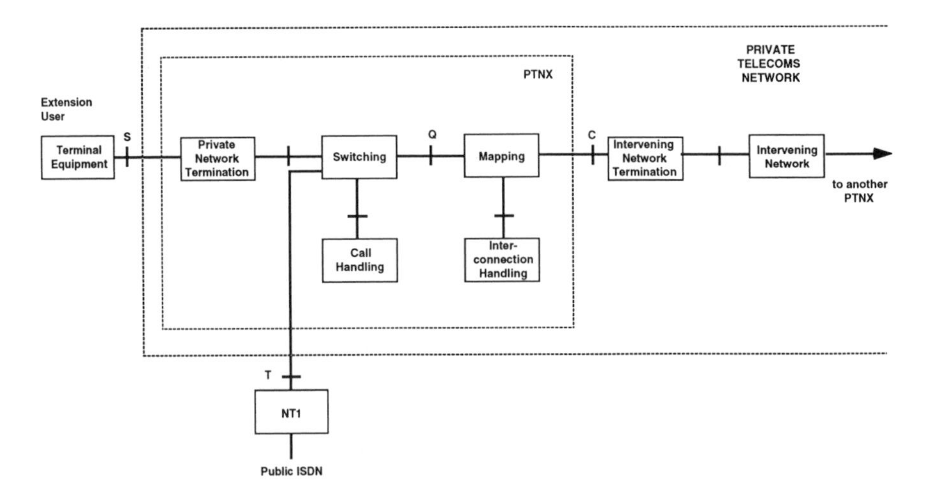

Figure 6.4
Outline of a physical implementation of a PT

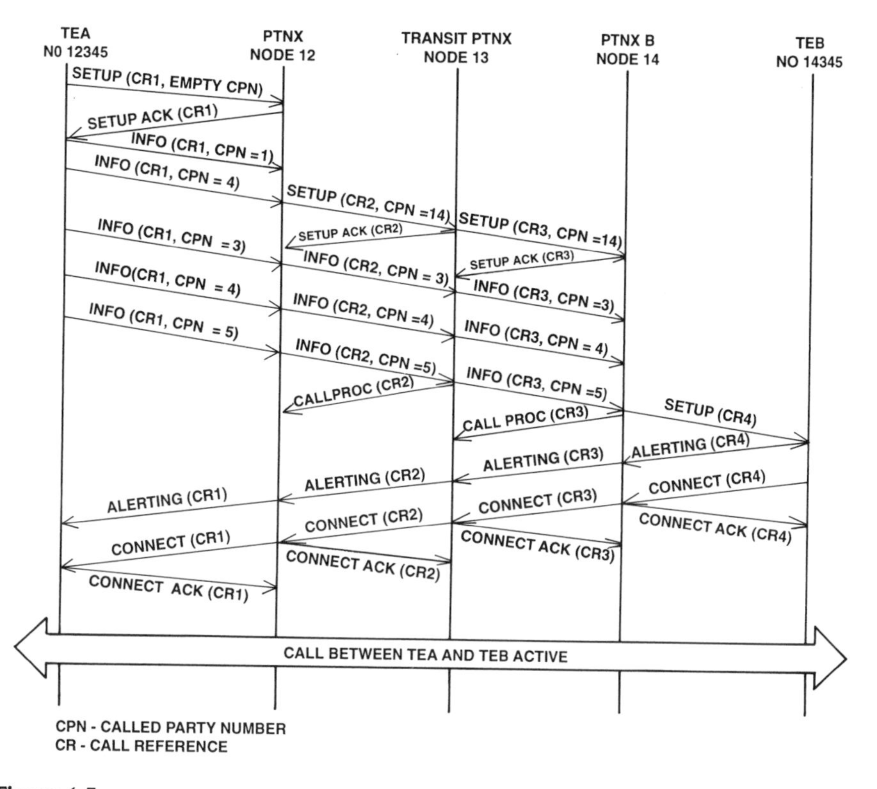

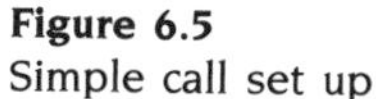

Figure 6.5
Simple call set up

the same messages between PTNXs as between terminals and the exchange. However, the most important feature of QSig (and DPNSS before it) is to allow the provision of facilities from a network of PTNXs as if all terminals were connected to a single PTNX even though the PTNXs may have been purchased from a wide range of manufacturers. For this purpose the message flows for a wide range of facilities are defined. Table 6.1 lists an initial range of facilities. It is not expected that every manufacturer's PTNX will be able to support all of these facilities. In the event of a user on one PTNX invoking a service which is not available on another manufacturer's PTNX, then a sensible fallback option is automatically offered where possible.

You will have noted that Standards can be itinerant. Many I series Standards actually originated elsewhere; similarly these ECMA Standards are on the move to ETSI and thence to the International Standards Organization (ISO). ISO introduces further nomenclature with its Private Signalling System (PSS1) for Private Integrated Services Networking (PISN) linking Private Integrated Network Exchanges (PINX).

Table 6.1
Q Sig features.

Calling Line Identification Presentation (SS–CLIP)	Do Not Disturb Override (SS–DNDO)
Connected Line Identification Presentation (SS–COLP)	Serial Call (SS–SC)
Calling/Connected Line Identification Restriction (SS-CLIR)	Completion of Calls to Busy Subscriber (SS–CCBS)
Multiple Subscriber Number (SS-MSN)	Completion of Calls on No Reply (SS–CCNR)
Calling Name Identification Presentation (SS–CNIP)	Call Waiting (SS–CW)
Connected Name Identification Presentation (SS–CONP)	Call Offer (SS–CO)
Calling/Connected Name Idenitification Restriction (SS–CNIR)	Intrusion (SS–INTR)
Malicious Call Identification (SS–MCID)	Terminal Portability (SS–TP)
Call Forwarding Unconditional (SS–CFU)	Call Hold ((SS–HOLD)
Call Forwarding on Busy (SS–CFB)	Explicit Call Transfer (SS–ECT)
Call Forwarding on No Reply (SS–CFNR)	Add-On Conference (SS–CONF)
Call Deflection (SS–CD)	Closed User Group (SS–CUG)
Controlled Diversion (SS–CDIV)	Call Pick-Up (SS–CPU)
Night Service (SS–NS)	Call Park (SS–CPK)
Network Interception (SS–NI)	Advice of Charge (SS–AOC)
Do Not Disturb (SS–DND)	User-to-User Signalling (SS–UUS)
	In-Call Modification (SS–IM)
	User Status (SS–UST)
	Multi-Level Precedence and Preemption (SS–MLPP)

REFERENCES

CCITT Recommendations:

I.421 Primary rate user–network interface.

I.431 Primary rate user–network interface—Layer 1 specification.

I.451(Q.931) ISDN user–network interface—Layer 3 specification for basic call control.

British Telecom Network Requirement (BTNR) 190. DASS2.

British Telecom Network Requirement (BTNR) 188. DPNSS.

In the USA—ANSI Standards:

TI.403 Carrier to customer installation—DS1 metallic interface.

TI.408 Primary rate user—Layer 1 specification.

TI.603 Minimal set of bearer services for the primary rate interface.

ECMA Standards:

ECMA-133	Reference configurations for calls through exchanges of Private Telecommunication Networks.
ECMA-134	Method for the specification of basic and supplementary services of Private Telecommunication Networks.
ECMA-155	Addressing in Private Telecommunication Networks.
ECMA-135	Scenarios for interconnections between exchanges of Private Telecommunication Networks.
ECMA-104	Physical layer at the interface of the Primary rate access extension line in Private Telecommunications Networks.
ECMA-123	In band parameter exchange in private pre-ISDN networks using Standard ECMA-102.
ECMA-105	Data link layer protocol for the D channel of the interfaces at the reference point between terminal equipment and Private Telecommunication Networks.
ECMA-141	Data link layer protocol at the Q reference point for the signalling channel between two Private Telecommunication Network exchanges.
ECMA-142	Specification, functional model and information flows for control aspects of circuit mode basic services in Private Telecommunication Networks.
ECMA-143	Layer 3 protocol for signalling between exchanges of Private Telecommunication Networks for the control of circuit-switched calls.
ECMA-106	Layer 3 protocol for signalling over the D channel of interfaces at the S reference point between Terminal Equipment and Private Telecommunication Networks for the control of circuit-switched calls.
ECMA-156	Generic stimulus procedure for the control of supplementary services using the keypad protocol at the S reference point.
ECMA-148	Identification supplementary services in Private Telecommunication Networks; specification, functional models and information flows.
ECMA-157	Protocol for signalling over the D channel of interfaces at the S reference point between Terminal Equipment and Private Telecommunication Networks for the support of identification supplementary services.

QUESTIONS

1 The European and North American network digital transmission systems carry 30 and 24 channels respectively. However the primary rate interfaces offered to the

customer, whilst remaining at 30 channels in Europe, drop to 23 channels in North America. Why the difference?

2 Why is it desirable to have compatibility between user–network signalling and inter-PABX signalling, and what are the different needs?

Chapter 7

Frame Mode Services

John Atkins

Early demand for switched data services, during the 1960s and early 1970s, arose largely from the high cost of computers. To make most effective use of the expensive data processing equipment, time-sharing by remote terminals became an attractive choice for many users. The ubiquitous public switched telephone network (PSTN) was pressed into service to provide the necessary switched access. As computer technology developed, however, it became clear that the PSTN would not be adequate for many of the new applications and that dedicated switched data networks would be required.

In 1967 Donald Davies at the National Physical Laboratories in the UK and Larry Roberts of the Department of Defense in the USA independently proposed packet switching as the most economical technique for switched data networks. The basic ideas for packet switching had been developed several years earlier by Baran and co-workers at the Rand Corporation to meet military requirements, but their work was shelved and apparently forgotten.

7.1 STORE-AND-FORWARD SWITCHING

Communication between computers or between computers and terminals usually involves the transfer of 'blocks' of data. Packet switching exploits the idea that data blocks may be transferred between terminals without setting up a continuous end-to-end connection. Instead they are transmitted on a link-by-link basis, being stored temporarily at each switch *en route* where they are queued for transmission on an appropriate outgoing link. Routing decisions are based on control information contained in a 'header' prefixing each data blocks. The term 'packet' refers to the header plus data block.

In fact the idea of switching by storing and forwarding is older than packet switching. It had been used in a variety of forms to provide message-switched services in which users could exchange complete messages—often very long messages—with the advantages of delayed delivery and broadcast options and retransmission if the message was garbled or lost in transmission. The distinctive feature of packet switching is that the packets, and consequently the queueing delays, are sufficiently short to permit interactive transactions.

7.2 DATAGRAMS AND VIRTUAL CIRCUITS

The simplest form of packet-switched service is the 'datagram' mode in which each packet is regarded as a complete transaction in itself. The network provides an independent switched path for each datagram. This implies that the header contains a complete set of information to route the datagram packet and that the network will have no pre-awareness of the connection.

Many transactions, however, involve the transfer of sequences of packets for which the 'virtual call' mode is more appropriate. In this case a virtual circuit is established by an initial exchange of signalling packets between the communicating terminals and the network. Full addresses are only used for this initial call set-up phase. The setting of a virtual call has the advantage that headers may be simpler, only indicating the virtual call identity. The disadvantage is that each node has to keep records of the routing relating to the call. The existence of two types of packet (signalling and data) means that each packet must be inspected to see whether it is for network control or for the customer. Data transfer then takes place and at the end of the transaction the call is cleared. During the data transfer phase the network tries to create the illusion of a real 'connection' whilst achieving the economies of dynamically shared resources.

Although the virtual call mode of operation is in general use, the debate continues as to the merits of the two manners of operation.

7.3 FLOW AND CONGESTION CONTROL

In a packet-switched network packets compete dynamically for the network resources (i.e. buffer storage, processing power, transmission capacity). A switch accepts a packet from a terminal largely in ignorance of what network resources will be available to handle it. There is therefore the possibility that a network will admit more traffic than it can carry, with a corresponding degradation in service. Controls are therefore needed to ensure that such congestion does not happen too often and that the network recovers gracefully when it does.

The importance of good flow control and congestion control is illustrated in Figure 7.1. For a given delay performance a packet-switched network may be regarded as having a notional maximum throughput, T packets/second. An ideal network would accept all traffic up to its throughput limit; beyond this some traffic would be rejected but the network's throughput would remain high (curve A). In a network with poor flow and congestion control, however, the throughput, whilst initially increasing with offered traffic, would eventually fall rapidly as a result of congestion (curve B). In practice good flow and congestion control maintain high throughput whilst leaving an adequate margin of safety against network

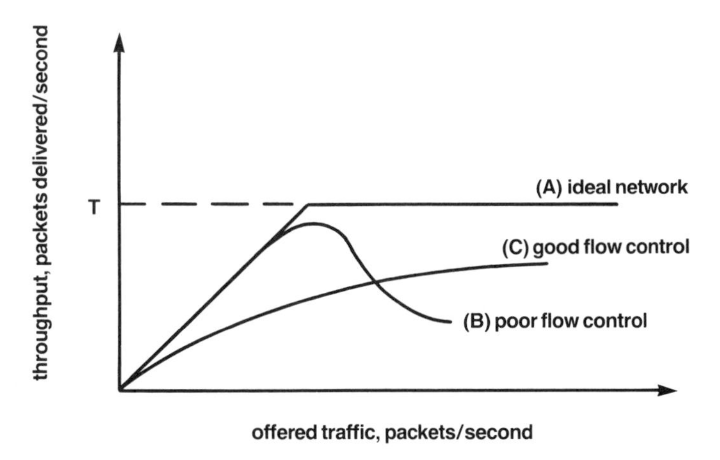

Figure 7.1
The importance of flow control.

failures and the imperfect operation of practical algorithms (curve C). Any flow control imposed by the network is in addition to end-to-end flow control applied by the terminals to match differences in speed. In fact end-to-end flow control is itself effective in avoiding network congestion since any increase in cross-network delay caused by the onset of congestion will delay the return to the sending terminal of permission-to-send, effectively reducing the rate at which it may send further packets.

The design of flow and congestion controls that are both effective and fair is difficult and remains the subject of active research. In general, flow and congestion control specify rules for the allocation and reservation of buffer storage and the selective rejection of incoming packets by switches when certain allocation thresholds (e.g. queue lengths) are exceeded.

7.4 STANDARDS

Growing interest in packet switching stimulated vigorous activity in CCITT, and in 1976 an international standard was agreed for packet-switched services. This is Recommendation X.25, which has provided the foundation for the development of Packet-Switched Public Data Networks (PSPDNs) around the world and has shaped the thinking of a generation of switching engineers. It is a virtual circuit protocol, structured after the ISO OSI 7-layer model in three layers, as shown in Figure 7.2.

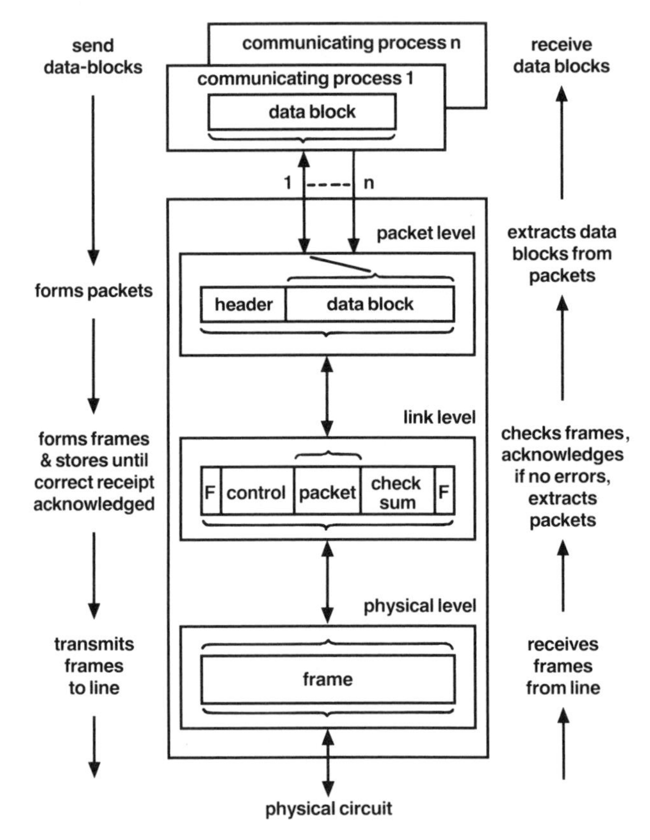

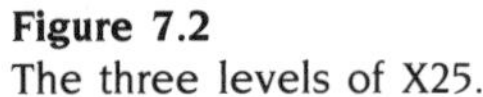

Figure 7.2
The three levels of X25.

7.5 SUPPORT OF X.25 BY AN ISDN

The ISDN arrived in an environment in which PSPDNs already existed. During the early years therefore ISDNs will depend largely on these separate PSPDNs to support packet mode bearer services. A comprehensive standard has been developed by CCITT which sets up arrangements for interworking between an ISDN and a PSPDN. This standard, CCITT Recommendation X.31 (also designated I.462), takes X.25 as the standard packet layer protocol, either in an X.25 terminal with a terminal adapter or in conjunction with the I.420 user–network interface to form a packet mode terminal equipment. In this case, the ISDN simply provides a transparent 64 kbit/s circuit between the packet terminal and a packet port on the associated PSPDN. This access circuit may be a semi-permanent or switched on demand B channel (Figure 7.3).

In addition to access to the services of a PSPDN, X.31 also defines what are called ISDN Virtual Circuit Bearer Services, based on X.25 packet handling capability contained within the ISDN itself. Access to the packet handler may be obtained either via a user's B channel or via the D channel as shown in Figure 7.4, and again may be either semi-permanent or switched on demand. The packet handler provides interworking with a PSPDN. This interworking may use the standard protocol X.75 which is designed for inter-network use, or may use a proprietary equivalent.

Depending on the application, packet access via the D channel may offer a number of distinct advantages, including:

—simultaneous operation of up to eight terminals on the ISDN passive bus, each terminal using a different data link connection identifier;

—unlike the B channels the D channel is essentially non-blocking;

—it leaves the B channels available for other services.

But D channel access also imposes some service limitations. The 16 kbit/s basic access D channel cannot support higher data rates and LAPD (see

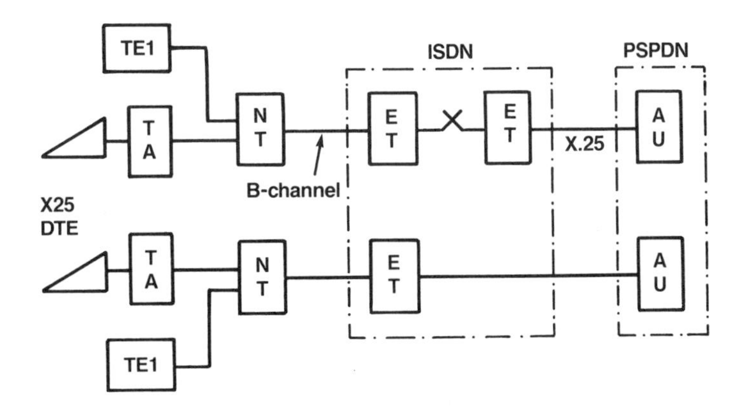

Figure 7.3
Access to PSPDN services.

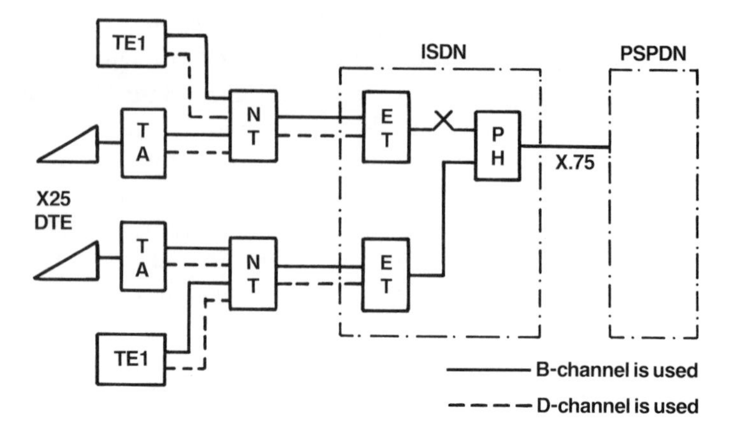

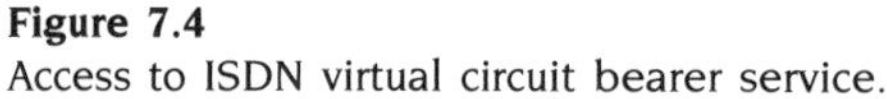

Figure 7.4
Access to ISDN virtual circuit bearer service.

Section 5.2) restricts X.25 packets to a maximum of 260 octets compared with 1024 for X.25. Furthermore, the additional delay caused by the slower D channel may be significant in some applications.

7.6 NEW ISDN FRAME MODE SERVICES

The term 'data' is traditionally associated with terminal-to-computer or computer-to-computer communications. But as new applications emerge and penetrate the workplace (and in due course the home) the scope for 'data' services is growing rapidly and it is becoming more appropriate to talk of non-voice services rather than data. Indeed, with the introduction of integrated multi-media applications even the distinction between voice and non-voice may become spurious.

Many of the new applications are 'bursty' in their requirement for transmission capacity (e.g. variable-bit-rate video) or involve short-duration transactions. Applications differ also in their sensitivity to delay or variations in delay, and in their ability to tolerate packets lost or corrupted because of transmission errors or congestion. There is also a general trend towards higher data rates (e.g. for video applications or LAN–LAN interconnection).

X.25 was developed specifically as a robust packet transfer service for 'traditional' data applications, but its complexity and lack of transparency (due to timeouts and flow control interactions) mean that it cannot support many of the new applications. It is equally clear that the ISDN should support new forms of frame mode bearer service that are able to offer the necessary flexibility. To this end a new family of ISDN frame mode bearer services is defined by CCITT.

One of the key features of the ISDN is the clear logical separation of signalling information and data. In CCITT parlance signalling belongs to the control plane or C-plane, whilst users' information belongs to the user plane or U-plane. In effect the ISDN may be viewed as two logically separate subnets—a signalling subnet and a switched information subnet, as shown in Figure 7.5. The new ISDN frame mode bearer services follow this established ISDN principle. Virtual calls are set up and controlled using the same out-band ISDN signalling procedures as circuit mode connections, unlike X.25 which carries signalling and data inseparably in the same logical channel.

Two distinct ISDN frame mode bearer services have been defined: frame relaying and frame switching. The same signalling procedures are used for both (Q.933), but they differ in the protocol supported by the network in the user-plane during the data transfer phase. They use data transfer protocols based on an enhancement of the standard LAPD Layer 2 signalling protocol known as LAPF—Link Access Procedures for Frame Mode Bearer Services (Q.922). A user may set up a number of virtual circuits and/or permanent virtual circuits simultaneously to different destinations.

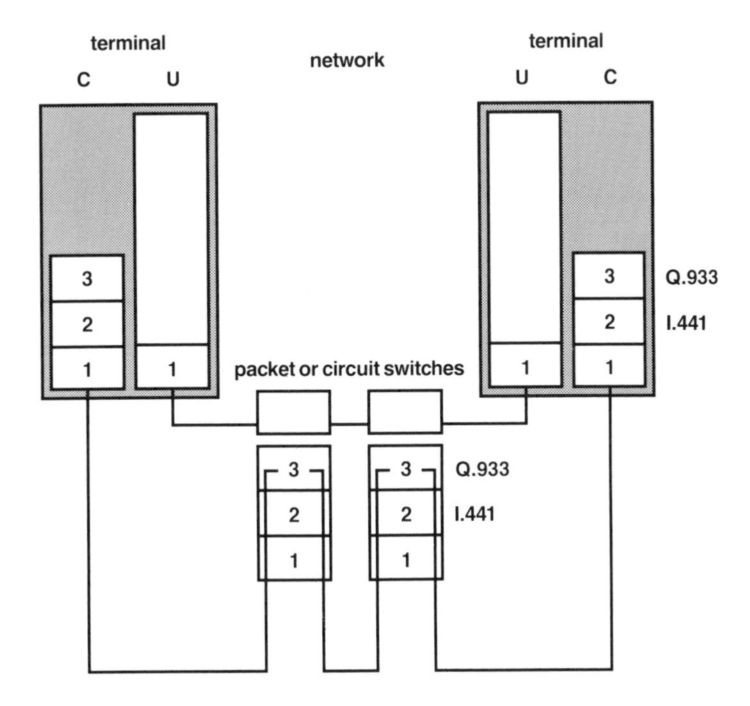

Figure 7.5
ISDN switched mode services.

Frame relaying

Figure 7.6 shows the protocol architecture for frame relaying, the simplest of the new frame mode services. Signalling in the C-plane uses the established LAPD signalling procedures at Layer 2 (Q.921 or I.441) with Layer 3 enhanced to support frame mode requirements (Q.933). But in the user plane the network supports only a very small part of the link layer protocol, generally referred to as the 'core' functions. These are:

—separating the frames using HDLC flags;

—forcing data transparency by zero bit insertion and extraction;

—checking frames for transmission errors and legth—any frames found to be in error are discarded;

—multiplexing and demultiplexing the frames associated with different virtual calls on the basis of their Layer 2 address. Many virtual calls may be supported simultaneously to different destinations.

The basic service provided by frame relaying is the unacknowledged transfer of frames between ISDN terminals. At its simplest the network creates an entry in a routing look-up table at call set-up time using the signalling procedures, a Layer 2 address being allocated to each virtual call

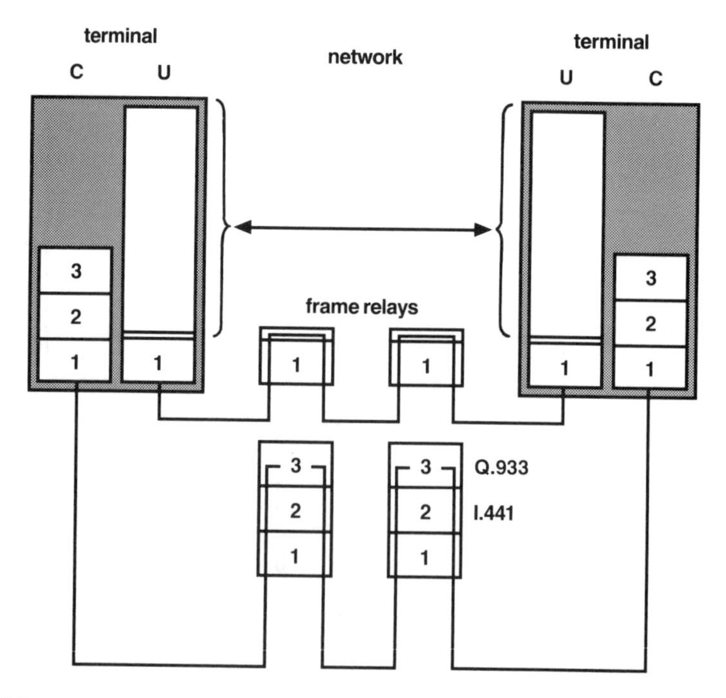

Figure 7.6
Frame relay frame mode.

for its duration. In the data transfer phase the network simply indexes the routing table using the Layer 2 address of the incoming frame, and queues it for transmission on the appropriate outgoing route.

In the data transfer phase terminals operate higher layer protocols on an end-to-end basis without involving the network.

Frame switching

The protocol architecture for frame switching is shown in Figure 7.7. Again signalling in the control-plane follows standard ISDN procedures but in this case the network operates the complete link layer protocol in the user plane during data transfer. Frame switching therefore provides acknowledged transfer of frames between terminals. The network detects and recovers from lost or duplicated frames and frames with errors and it operates network-enforced flow control. As with frame relaying, multiplexing, demultiplexing, and routing at each switching node is done on the basis of Layer 2 address. Again, higher layer protocols are operated by the terminals on an end-to-end basis.

The new ISDN frame mode bearer services represent a significant extension to the capabilities of the emerging ISDN. The architectural

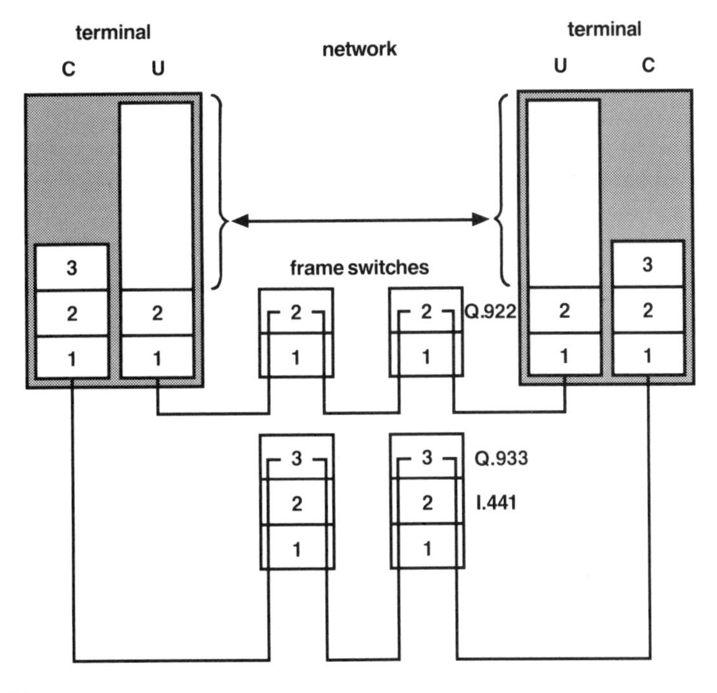

Figure 7.7
Frame switching frame mode.

integration of frame mode services with the complementary circuit mode services will provide the flexibility needed to meet the growing range of users' requirements. The lightweight data transfer protocol of frame relaying will support applications which cannot tolerate the additional delays that would be caused by error-correction-by-retransmission. And it means that frame relaying will provide a high degree of protocol transparency, there being a minimum of interaction with higher layer protocols. Frame switching on the other hand will offer the robustness against errors usually associated with X.25 but without the per-packet overhead of Layer 3.

7.7 FRAME FORMAT

In the data transfer phase of frame relaying, information is carried in frames conforming to the format shown in Figure 7.8.

Figure 7.8
Frame structure

The address field contains a minimum of two octets but may optionally be extended to four octets.

The information field may be as small as a single octet. The default maximum length of this field is 262 octets, which is compatible with LAPD on the D channel (the LAPD frame contains a control field of up to 2 octets and a maximum of 260 octets in the information field). Other maximum information field lengths may be negotiated at virtual call set-up time. It is recommended that networks support a negotiated maximum field length of at least 1600 octets to avoid the segmentation and reassembly that may otherwise occur, e.g. in LAN interconnection applications.

REFERENCES

CCITT Recommendations:

The Standard adopted for most public packet-switched networks is:

X.25 Interface between data terminal equipment and data circuit terminating equipment for terminals operating in a packet mode and connected to public data networks by dedicated circuit.

There are three other Standards relating to the connection of terminals, often referred to as the 'triple-X' protocol:

X.3 Packet assembly/disassembly facility (PAD) in the public data network.

X.28 DTE/DCE interface for a start–stop mode data terminal equipment accessing the packet assembly/disassembly facility (PAD) in a public data network situated in the same country.

X.29 Procedures for the exchange of control information and user data between a packet assembly/disassembly (PAD) facility and a packet mode DTE or another PAD.

X.25 on the ISDN is described in:

X.31 Support of packet mode terminal equipment by the ISDN.

The internetwork protocol is:

X.75 Terminal and transit call control procedures and data transfer system on international circuits between packet-switched data networks.

New ISDN frame mode services:

I.122 Framework for providing frame mode bearer services.

I.232 Frame mode bearer service categories.

I.233.1 ISDN Frame Mode Bearer Services (FMBS)—ISDN Frame Relaying Bearer Service.

I.233.2 ISDN Frame Mode Bearer Services (FMBS)—ISDN Frame Switching Bearer Service.

I.370 Congestion Management for the ISDN Frame Relaying Bearer Service.

Q.922 ISDN Data Link Layer Specification for Frame Mode Bearer Services.

Q.933 Digital Subscriber Signalling System No 1 (DSS1)—Signalling Modifications for Frame Mode Bearer Services.

In the USA – ANSI Standards:

TI.606 Frame Relay Service Description.

TI.608 Packet Mode Bearer Services. Control procedures—BRI and PRI.

TI.617 Signalling Specification for frame relay bearer services.

TI.618 Core Aspects of Frame Relay Protocol.

A general book on the subject is *Fast Packet and Frame Relay* by Rob Schweidler and Jim Marggraff published by FH Publication.

QUESTIONS

1 When X.25 was introduced it made a feature of the fact that signalling and data were treated in a common manner. Frame mode services now make a feature of the separation of signalling and user data. Explain the apparent contradiction.

2 Packet and frame mode services can operate in both the B channel and the D channel. What are the merits of the B channel compared with the D channel for this purpose?

3 Figure 7.1 implies that in some circumstances a network with 'poor' flow control can carry more traffic than one with 'good' flow control. Why is this?

Chapter 8

ISDN Customer Premises Equipment and Applications

The ISDN as described so far has only considered the provision of control and traffic channels for customer's use. In CCITT terms this is equivalent to an ISDN bearer service, defined as follows:

> A provision of such a service to the customer allows him the capability for information transfer between ISDN access points according to CCITT standards. The higher layer (i.e. above Layer 3) terminal functions are defined by the customer.

That is to say the customer is free to make whatever use of the 64 kbit/s bearer he or she may wish, without constraint. However, in practical terms such liberalization leads to anarchy rather than communications. It would be farcical, for example, for every manufacturer to choose their own technique of 64 kbit/s speech encoding so that only customers using equipments of the same manufacture could speak to one another. Nevertheless such situations have arisen in the past between facsimile equipments and other video and data exchange equipment. To make maximum use of the bearers it is necessary that the higher layers of the customer-to-customer protocol are defined so that interworking is possible. Such a defined service is known by CCITT as a 'teleservice', defined as:

> This type of service provides the user with the complete capability, including defined terminal equipment functions, for communication with another user of the service according to protocols established by CCITT.

The intention of the CCITT in this categorization is to distinguish services according to what technical Standards are needed for their specification and implementation such that international interconnection is possible. It was not the intention to categorize services according to different possible arrangements of equipment ownership or the provision to the customer of the means to support the services.

By examining the functional picture of the access arrangement to the ISDN (Figure 8.1), it is possible to have a clearer understanding of the differences and the relationship between these two categories of services.

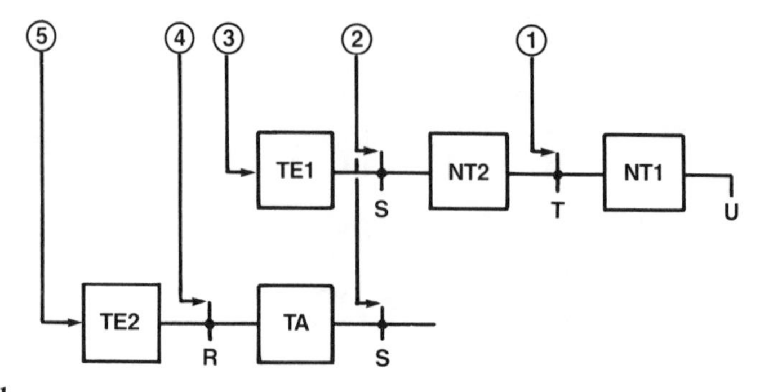

Figure 8.1
Customer access to services supported by an ISDN.

Access points 1 and 2 are the access points for the bearer services supported by an ISDN. At access point 4, depending on the types of terminal adapters provided, other CCITT standardized services may be accessed. These could be called dedicated network bearer services such as the circuit-switched or packet-switched data services associated with user-network interfaces according to Recommendations X.21 or X.25 respectively.

Access points 3 and 5 are the access points for teleservices. In this Chapter we shall look at the applications of ISDN. In particular we shall highlight those applications which particularly benefit from the availability of 64 kbit/s bearers.

Services can benefit at three levels from the availability of a 64 kbit/s bearer.

1. Those which benefit in a quantitative way but whose principles are basically the same as when operating at lower speeds. The obvious examples are file transfer and electronic mail. Changing from 9.6 kbit/s modem based bearers to 64 kbit/s bearers means that the transfer rate is increased almost seven-fold and times correspondingly reduce. Nevertheless the principles are the same. Note that it may be that the seven-fold improvement is not readily achievable because error control protocols, which are optimized for the lower data rates, may be less than optimal at higher speeds owing to software limitations, and because the number of bits delay in the link is much higher on the ISDN connection both due to the higher bit rate and the switching mechanisms employed.
2. Those which benefit to such a degree that totally new uses of the application appear. Examples are:

   (a) Facsimile transmission is obviously possible over the analogue network, but the 20 or 30 seconds per page means that it is not an attractive proposition for more than a few pages. On ISDN, at 64 kbit/s, pages can be transmitted in 3 or 4 seconds and this moves the processes into the realm of the 'slow photocopier' where tens of pages could be transmitted with little effort. It is arguable that in practice the ISDN facsimile service will be applied entirely differently to the analogue service.

(b) Videotex services such as Prestel (UK), Minitel (France), Bildschirmtext (Germany) will be very different services when they are moved from 1200 bit/s to 64 kbit/s. Pages can be written instantly and browsing is possible. Colour pictures can be included in the pages which adds a whole new dimension for marketing and information distribution.

3. Those whose provision is totally dependent on the availability of 64 kbit/s links, which are not realizable at lesser data rates or on analogue bearers. Particular examples:

   (a) Videophone whose performance may be acceptable at 64 kbit/s, which at lower rates does not at present appear usable.

   (b) High quality speech which uses a coding algorithm which is more complex than PCM to convey better quality speech, or even music. No equivalent service is possible over analogue networks.

As discussed earlier in the chapter, such applications must be standardized so that terminals from a wide range of sources can interwork. In the next sections the operation of systems in categories 2 and 3 above will be discussed.

8.1 HIGH QUALITY SPEECH

Paul Challener

The process of encoding speech into PCM is described in Chapter 2. PCM is a simple technique consisting of filtering to avoid the overlapping of sidebands in the sampling process (known as aliasing), sampling at an 8 kHz sampling rate and 256-level quantization. At each sampling instant, the quantizer outputs an 8-bit codeword corresponding to the quantization level chosen. Hence the overall bit-rate is 64 kbit/s. The coded signal bandwidth is 300 Hz to 3.4 kHz which results from the anti-alias filter. This may not be seen as particularly speech specific, but the design of the quantizer does reflect understanding of the speech signal. The 8-bit codewords result from a 12-bit to 8-bit compression following a logarithmic law (A-law in Europe and μ-law in USA), such that the quantization noise encountered on coding large signal segments is generally larger than that encountered when coding small signals. Thus the signal-to-noise ratio remains approximately independent of the signal level. The advantage of this technique for coding speech is that the signal hides most of the quantization noise so that it is not perceived by the listener; this is known as 'auditory masking'.

Algorithms have been developed which exploit redundancies in the speech signal more than PCM, resulting in further reductions in the bit-rate necessary to code telephone quality speech. In the last few years such techniques have also been applied to wide bandwidth, high quality speech signals in an attempt to code them into the standard 64 kbit/s channel of ISDN. This has reached acceptance in a CCITT Standard G.722 which covers conversion of a 7 kHz bandwidth speech signal into a 64 kbit/s bearer. The rest of this section is intended to provide an understanding of this Standard with an insight into the techniques employed and the properties of speech which are exploited in achieving them. However, as the workings of a technique known as Adaptive Differential Pulse Code Modulation (ADPCM) are central to an understanding of the Standard, this will be described first.

8.1.1 Speech coding using ADPCM

A major step forward in coding speech more efficiently than PCM came in 1984 when ADPCM was accepted by the CCITT for coding telephone quality speech at 32 kbit/s (G.721). In this Standard the target quality was identical to that of 64 kbit/s PCM on a subjective level. In other words, it would not matter if there was a measurable decrease in signal-to-noise ratio if this could not be perceived by the listener.

The principal differences between PCM and ADPCM are contained in the 'A' and the 'D' which stand for 'adaptive quantization' and 'differential encoding' respectively.

An 'adaptive quantizer' differs from the standard PCM quantizer in two significant ways. First, at any given sampling instant, all the quantization levels are uniformly spaced with an interval S; they are not logarithmically spaced. Secondly, when signals are quantized towards the limits of the quantization range the step size S is increased; when they are quantized towards the centre of the range, S is decreased; elsewhere the step size hardly alters. As speech signals from a single talker do not change in level significantly from one word to the next, this kind of quantizer can quickly adapt to a range suitable for the actual signal conditions and then remain almost perfectly scaled until the speaker stops talking. Figure 8.2 illustrates the operation of the adaptive quantizer. Speech which is initially linearly encoded into 12 bits (i.e. 4096 levels) is then compared with 16 quantizing

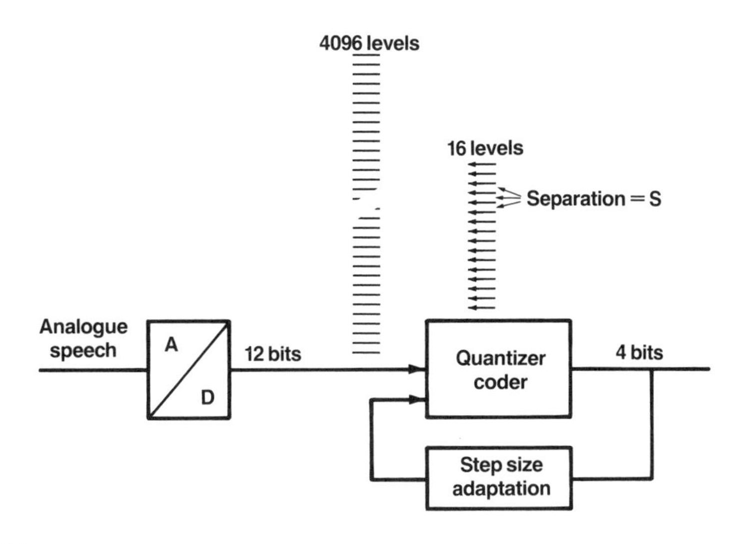

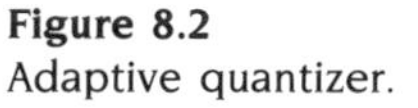

Figure 8.2
Adaptive quantizer.

levels spaced S apart and the nearest quantization level is chosen to represent the speech. As there are 16 levels, the chosen level can be encoded as 4 bits.

The expansion or compression of the quantization range is achieved by simply multiplying S by a number slightly greater than unity or slightly less than unity dependent only upon the quantization level selected to code the previous signal sample. Such a dependence only upon past events to determine current values is termed a 'backward adaptive process'. The reason for this name can easily be seen from Figure 8.2 where the direction of the step size adaptation process opposes the normal signal flow through the coder.

Backwards adaptation is an elegant technique which can be simultaneously performed at the encoder and decoder without the need for external side information to keep the processes in step; they are each driven from the output of the coder. Furthermore, it is possible to include the concept of 'leakage' in the quantizer to improve its robustness to channel errors; this means that in addition to the control on S exerted by the output of the decoder, S is also multiplied by an additional factor to cause it to tend to a fixed value. This ensures that departures from synchronization are leaked away to zero. In this way, the encode and decode processes will automatically align themselves at start-up or after bursts of errors have corrupted the transmitted signal values. Hence all the signal bits fed to line are truly used to encode speech efficiently, not to control terminal processes, and the result is a good mixture of simplicity, robustness and elegant design.

'Differential encoding' implies that the quantizer input is not obtained from the signal itself, but from the difference between current signal values and previous ones. In ADPCM, this concept is progressed further by the inclusion of a predictor circuit in the differential loop. Figure 8.3 shows this; the outer loop implements the differential process. Predictors are adaptive filter circuits which use previous sample values to construct an estimate of the signal at the current sampling instant. When the real current value arrives at the coder input, the estimate is then subtracted from it leaving only a tiny difference signal which is fed to the quantizer. Hence, the differential process reduces the size of the signal fed to the quantizer and thus enables fewer bits to be used in the quantizer giving rise to a second significant bit-rate reduction over standard PCM.

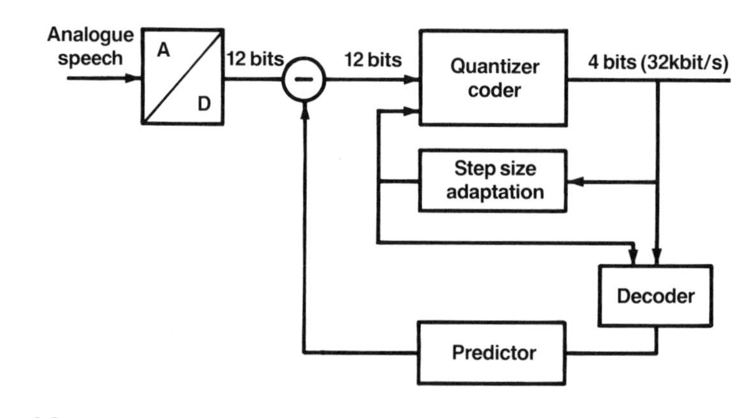

Figure 8.3
32 kbit/s ADPCM encoder.

The differential prediction process is also a backwards adaptive process which again requires no side information to keep the decoder in synchronism with the encoder. Clearly, the closer the predictor estimate is to the actual current signal sample, the smaller the size of the signal fed to the quantizer and hence the fewer bits needed to encode it. The theory of predictors is a difficult and demanding task which is outside the scope of this book. Suffice to say that some knowledge of the signal statistics is necessary in order to design the predictor filter appropriately. In the case of 32 kbit/s this knowledge is obtained by modelling the signal spectrum by a transfer function whose six poles and two zeros are determined by the long-term and short-term characteristics of the signal respectively. The property of speech being exploited is that it is highly predictable for much of the time; departures only occur at boundaries between the elements of speech which constitute real information.

8.1.2 Sub-band ADPCM for high quality speech

On ISDN, the B channel rates are 64 kbit/s and so there is little benefit in using 32 kbit/s ADPCM for single channel telephony applications. However, when higher quality audio is required, there is clearly some scope within the 64 kbit/s bit-rate to widen the bandwidth and improve the dynamic range outside the limits of 64 kbit/s PCM. However, before considering what techniques to use the first question is what signals do we wish to transmit?

There are obviously applications for systems which transmit compact disc quality audio (50 Hz to 20 kHz), but such a target would currently be impossible to achieve within a single 64 kbit/s B channel. However, in many possible ISDN applications such as conferencing or commentary broadcasting, the only audio signal is speech, and tremendous benefits can be achieved if this is coded to a subjectively high quality. So the next question to consider is what is the real bandwidth of speech?

The answer to this surprises many as it does not match the auditory bandwidth. Speech coding experts generally consider it to be about 50 Hz to 7 kHz. The low frequency limit is due mostly to the presence of low pitch frequencies in some voice speech, being particularly relevant in 'fruity' voiced male speech. The high frequency limit is largely due to limits on the mechanics of the human speech production equipment. It requires muscle control to move the tongue, lips and teeth to create plosives or stops (e.g. 'B' or 'T') and hence these do not contain very high audio frequencies. Only the fricatives (hissing sounds like SSS, FFF) contain energy towards the 7 kHz region and even then the power at these frequencies is small. Hence, if the spectrum from 50 Hz to 7 kHz can be faithfully reproduced we can consider we have achieved a high quality speech system. This was the target which the CCITT set itself in the late 1980s and the result was the Standard G.722. It was so successful that it proved possible to generate additional data channels alongside the audio with little or no apparent degradation to the speech signal. The operation of this standard is described below in conjunction with Figure 8.4.

Input speech is first sampled at 16 kHz and digitized to at least 14 bits before being fed into the first digital signal processing block marked 'Transmit QMF' in Figure 8.4. 'QMF' stands for Quadrature Mirror Filters and consists of a pair of digital bandpass filters which separate the signal into two bands, a low band and a high band, with frequency ranges 50 Hz to 4 kHz and 4 kHz to 7 kHz. They are not 'brickwall' filters but deliberatley overlap each other in the 4 kHz region in a predetermined manner. The two bands are then downsampled (by taking alternate samples) to 8 kHz as each contains less than 4 kHz signal bandwidth and the reduced sample rate signals are then coded separately. Some slight aliasing occurs because of the QMF filter overlap at 4 kHz but this will not ultimately cause

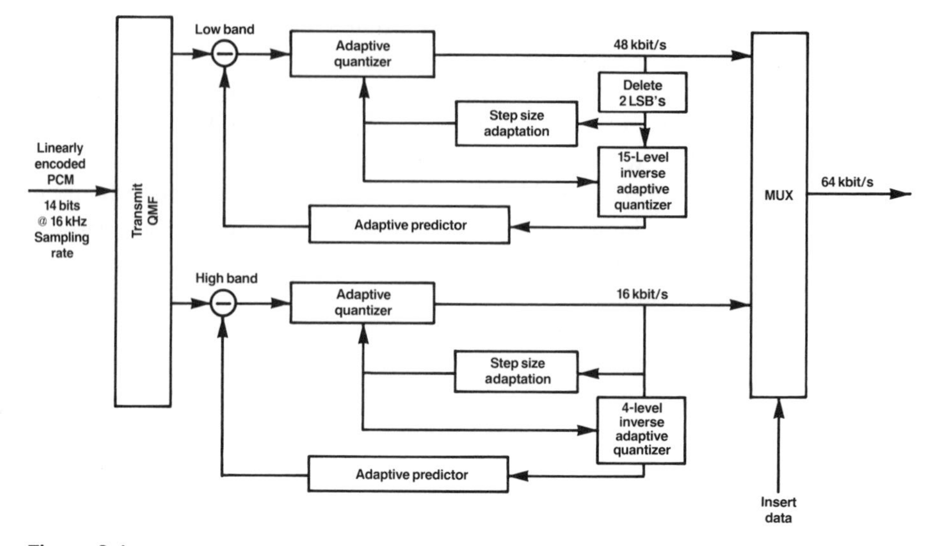

Figure 8.4
7 kHz sub-band ADPCM encoder.

concern. In each band, the coding process is ADPCM, but in the case of the low band, where most of the speech energy lies, 6 bits are allocated to the quantizer whereas for the high band only 2 bits are allocated. (Hence the 48 and 16 kbit/s data streams shown in Figure 8.4 for the two bands.) These two datastreams are combined in a multiplexer and sent to line at a composite 64 kbit/s rate.

In the decoder, the reverse processing occurs – demultiplexing of the two bands, decoding the ADPCM in each and then upsampling to a 16 kbit/s sampling rate and finally recombining through a receive QMF filter bank. Cleverly, in this final process, the alias terms which were introduced on downsampling are cancelled and the result is very high quality speech. An interesting fact is that this process could never have been performed with purely analogue processing; the degree of alias cancellation is a result of precise filter characteristics implemented to a tolerance beyond the dreams of analogue designers; they are also not subject to drift through ageing or temperature variation and are therefore truly a product of the digital era! Hence, in addition to the properties of speech exploited by pure ADPCM, sub-band ADPCM allows us to take account of the spectral shape of speech, allocating most bits to parts of the spectrum with most energy and less bits to those with least, without affecting the overall bit rate. Furthermore, any quantization noise which occurs due to quantizing a signal in one band will be filtered by the QMF filters so that it also lies in that band. In this way, the noise will tend to be well masked by the signal which caused it as they must be reasonably close in frequency. In Figure 8.4, it can be seen that the ADPCM encoder for the low band is more complex than that for the high band. The technique adopted to provide the data channel is to steal one or two of the 6 bits from the low band at each sample instant, giving rise to an 8 kbit/s or 16 kbit/s data channel. The multiplexer allows external data to be inserted when these channels are available and ensures appropriate interleaving in specific modes of operation. The trick employed here is that the low band ADPCM coder is what is called an 'embedded' coder. This implies that whether external data is present or not, the most significant bits of the speech low band signal are always present in the same positions in the multiplex. Only the least significant bits are dislodged to allow external data to be sent and then only for the time when external data is actually present; once the external data input ceases, all 6 bits per sample are given back to speech. In this way a flexible speech and data system is ensured and even if the multiplex mode is misaligned due to transmission errors, the received speech is only affected in least significant bits which are scarcely audible. The use of this data channel is discussed in Section 8.6.

8.2 MUSIC CODING

The coding system described in the last section was designed around the requirements for speech. At the time when its design was finalized some account was taken of its possible use for the transmission of music and in practice it is often used for that purpose. However, in a world used to compact discs its quality falls well below high fidelity. CD coding typically requires 16-bit sampling at 44.1 kHz giving a bit rate of 705 kbit/s; for a stereo signal twice this rate is required. For transmission and storage purposes such a bit-rate is inconveniently high and so considerable effort has been put into reducing the rate. In fact most of the initial drive came from those developing techniques for storing both audio and video images at rates up to 1.5 Mbit/s. The audio part of the system is popularly known as 'Musicam', although this is strictly the name of a precursor to the system defined by the ISO MPEG (Moving Picture Experts Group), now described in ISO standard 11172.

Musicam (Masking pattern Universal Sub-band Integrated Coding And Multiplexing) extends the principle of splitting the audio band into sub-bands. In Musicam coding an audio band of 20 kHz is split into 32 equal sub-bands. The bit-rate reduction process relies on the masking effects of the ear. In the presence of a high amplitude tone the ear is insensitive to lower amplitude signals of similar frequency. The closer is the frequency, the more is the masking effect as is shown in figure 8.5. Similarly the high level tone will mask quantizing distortion. By spitting the band into sub-bands the masking threshold due to the higher level signals can be

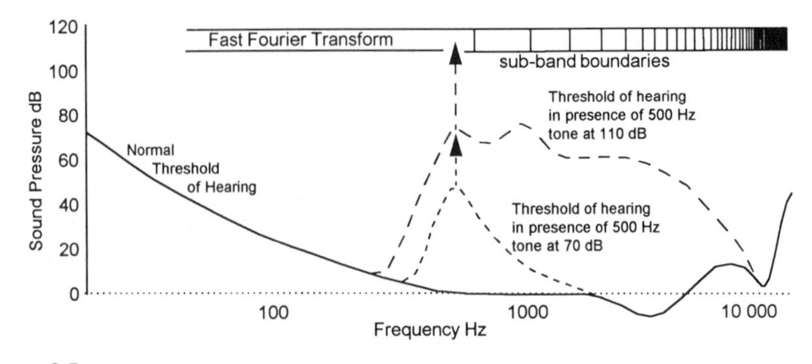

Figure 8.5
Musicam encoding process.

calculated for each band, and information transmitted relating only to those bands which contain information whose level exceeds the masking threshold. Where information relating to a band is transmitted it may be quantized efficiently so that a minimum number of bits is required, keeping the quantizing distortion just below the masking level.

At higher frequencies the masking levels may be determined by the filter outputs, but to improve the accuracy at lower frequencies a Fast Fourier Transform (FFT) is done in parallel with the sub-band filtering. For each of the sub-bands the scale factor is calculated based on the maximum amplitude in that band over a frame of samples. If the content of a band is above the masking level then the encoding samples and scale factors are multiplexed for transmission. The resulting bit-rate is not constant. Practical measurements show that the bit-rate required seldom exceeds 110 kbit/s; if less than this rate is available (e.g. a single B channel) then the multiplexer removes such information as will introduce the least audio distortion. Listeners report that at 128 kbit/s the quality is indistinguishable from a CD.

So far only monophonic channels have been considered. If stereo channels are to be transmitted then further economies in bandwidth can be made. Experiments show that in locating an acoustic image the fine spectral detail is important at low frequencies but only the level of the high frequency (above about 2 kHz) components is relevant. Hence the sum signal is encoded and transmitted for the high frequency bands together with the scale factor signals for both channels so that the intensity of each channel is preserved while the unimportant fine detail is lost. When transmitting a stereo signal, if the required bit-rate exceeds the available bit-rate then some of the higher frequency bands are encoded in this manner. This is known as joint stereo coding.

It will be appreciated that this description of the coding process is very outline in nature. Sampling of the audio signal may be at 32 kHz, 44.1 kHz or 48 kHz. The Standard allows for three layers of coding, the complexity increasing with the higher layers. Layer 1 covers the basic mapping into the 32 sub-bands, the segmenting of the signal into frames of 384 samples and the determination of scale factors. Layer 2 provides for a frame of 1152 samples with additional coding of bit allocation, scale factors and samples. Layer 3 allows the dynamic variation of the sub-bands, non-uniform quantization and Huffman coding. Any decoder can recognize and decode signals at its design level and levels below; for example a Layer 2 decoder can decode both Layer 2 and Layer 1 coded signals. Also included in the frame structure is a degree of error protection and the facility for a low speed data channel. There are also options in the psychoacoustic process for determining the masking level. This will affect the decisions as to which band's information will be transmitted; the choice has no effect on the decoding process as the decoder will base its output on the information supplied, however it is generated.

8.3 FACSIMILE

Malcolm Jones

Facsimile image transmission, a method of transmitting unstructured images using a raster-scanning technique, pre-dates telephony by several decades. It was first patented in 1843 by Alexander Bain as a development of the master–slave–clock time distribution system which provided a method of synchronizing two pendulums over a communications channel.

The image transmitting station scanned an electrically conducting image, such as a page of metallic type, with a stylus on the end of its pendulum and the receiving station applied ink to paper from the end of a synchronized pendulum.

In 1850 the pendulums were replaced by the cylinder and screw scanning mechanism and in 1902 the photo-electric scanning technique was patented by Dr Arthur Korn of Germany. By 1910 there was an international facsimile service between London, Paris and Berlin using Korn equipment.

Until the 1960s the market for facsimile equipment was small and specialized (e.g. phototelegraphy for the newspaper industry). In 1968 the CCITT produced a Recommendation for Group 1 facsimile equipment, as distinct from phototelegraphy, which defined an analogue FM transmission system for sending a grey-scale image of a page via the PSTN in 6 minutes at a vertical resolution of 3.85 scan lines per millimetre. In 1976 an improved system was defined which reduced the transmission time to 3 minutes using a special analogue VSB transmission technique. This was followed by the definition of the first digital form of facsimile, Group 3, in 1980.

Group 3 facsimile broke with the tradition of previous CCITT Standards by not defining a method of supporting grey-scale image transmission. The Group 3 image is first quantized on a pel-by-pel basis (pel = picture element) into one of only two 'colours', either black or white. This dichotomization allowed a run-length coded compression scheme to be applied to the picture data which resulted in a reduced page transmission time of 1 minute when using a 4800 bit/s communication system. Group 3 offers a digital horizontal pel resolution of 1728 pels per 215 mm or 8.03 pel/mm. The basic vertical scanline resolution is 3.85 line/mm as in the earlier groups but there is an optional higher vertical resolution of double the basic value, 7.7 line/mm.

The introduction of a compression technique into any system leads to some new problems. These are principally the magnification of the effect of communication errors and the possibility that the transmitter may overrun the receiver. To deal with the error extension problem in Group 3 the compression algorithm is designed to operate on only a very limited number of scanlines at a time and each compressed scan line is terminated with a long robust end-of-line code. The number of scanlines processed at any given instant is referred as the *k* factor and is basically 1 but may be negotiated between the terminals up to a maximum of 4. This small number is necessary because the original version of Group 3 did not include any error correction in its communication system. Despite this limitation the compression system is reasonably efficient giving compression factors between 7 : 1 and 12 : 1 on most documents.

The communication system for Group 3 facsimile operated at 4800 bit/s but optional speeds of 9600, 7200 and 2400 bit/s were also defined. When operating at 9600 bit/s a single A4 page image takes approximately 30 seconds to transmit depending upon the complexity of the image.

One factor which has contributed to the success of Group 3 was the introduction of the flat-bed scanner which replaced the old cylinder and screw mechanism and allowed multi-sheet scanners to be built into facsimile machines at reasonable cost.

The Group 3 facsimile terminal is one of the most successful telecommunication terminals ever devised with the number installed being several times greater than the telex terminal population, for example. Facsimile is now the second service on the PSTN after telephony and it is generating very large volumes of traffic particularly on international routes where the office hours at each end do not overlap. The figures for 1989 suggest that some 10 million terminals had been installed with a growth rate over 30% per annum and no sign of market saturation.

8.3.1 Group 4 facsimile for ISDN

When a new facsimile group was being discussed for use on the ISDN in the early 1980s this phenomenal success was not yet visible and a single solution was being sought for all kinds of non-voice traffic under the generic title 'Telematics'. This unified approach to telecommunications services matched the universal concept for communications protocol design as defined by the International Standards Organization (ISO) in their 'Open Systems Interconnect' (OSI) concept. Thus Group 4 facsimile was defined in 1984 as one of a family of services which also included basic teletex with its inherent access to the telex service, and mixed mode, a form of document coding which supported both character and facsimile image blocks within a single page. All these services used the same communications protocol stack and only differed at the application layer. Teletex and Group 4 facsimile had interworking arrangements included by creating three classes in the Group 4 definition.

Class 1 only supports basic facsimile without interworking to teletex.

Class 2 is compatible with class 1 but it is also capable of receiving basic teletex and mixed mode calls.

Class 3 is compatible with class 1 but it is also capable of both transmitting and receiving basic teletex and mixed mode calls.

All types of Telematic terminal employ a 7-layer communications protocol stack similar to the ISO's OSI.

8.3.2 Group 4 compression algorithm

The Group 4 facsimile system is fully digital in all respects including having an error-correcting communications protocol. Therefore it can employ a redundancy removing compression algorithm. The coding concept is based upon the identification of run lengths of the same colour between adjacent scan lines and is therefore referred to as two-dimensional coding or 2D coding. It is also called Modified Modified Reed (MMR) coding.

The coding of the run lengths is simply performed by selecting variable length Huffman codes. The principle of Huffman coding is to allocate short codes to those codewords which occur most often and longer codes to less frequently occurring codewords. If there is any redundancy in the original sequence then the resulting data is reduced compared with the original. For a sequence of known statistics an optimum code translation can be identified. In this case a pre-defined dictionary is used. The dictionary is made up of two parts. One part contains the short run lengths from 0 to 63 pels in steps of one pel which is referred to as the 'terminating codes table'

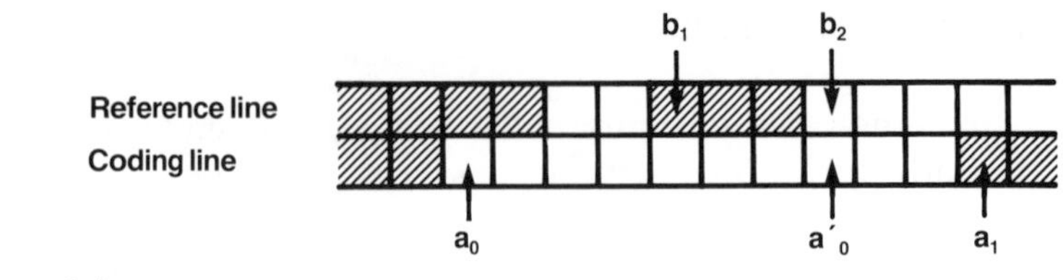

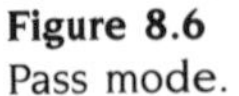

Figure 8.6
Pass mode.

and the second part contains the make-up codes for run lengths from 64 to a full scanline of 1728 pels, in the A4 paper case, in increments of 64 pels. The dictionary contents are pre-defined and fixed. There is no dynamic up-dating process.

Identification of the run lengths is performed in three different ways depending upon the precise pel pattern and its position on the page. Note that facsimile scanners overscan the paper, to ensure that all the image is captured, and therefore all scan lines start with a white interval plus the first scan line is an imaginary all-white line.

The three coding modes are referred to as:

(a) *Pass mode* (see Figure 8.6). This mode is identified when the position of b_2 lies to the left of a_1. However, the state where b_2 occurs just above a_1 is not considered as a pass mode.

(b) *Vertical mode* (see Figure 8.7). When this mode is identified, the position of a_1 is coded relative to the position of b_1. The relative distance a_1b_1 can take on one of seven values $V(0)$, $V_R(1)$, $V_R(2)$, $V_R(3)$, $V_L(1)$, $V_L(2)$ and $V_L(3)$, each of which is represented by a separate codeword. The subscripts R and L indicate that a_1 is to the right or left respectively of b_1 and the number in brackets indicates the value of the distance a_1b_1.

(c) *Horizontal mode*. When this mode is identified, both the run lengths a_0a_1 and a_1a_2 are coded using the codewords $H + M(a_0a_1) + M(a_1a_2)$. H is the flag code word 001 taken from the two-dimensional code table. $M(a_0a_1)$

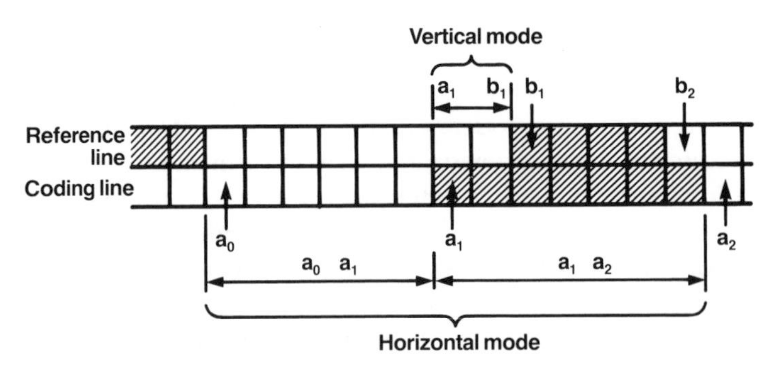

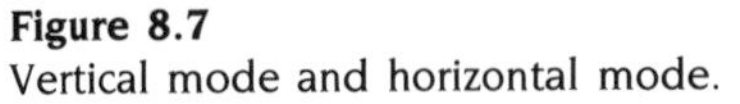

Figure 8.7
Vertical mode and horizontal mode.

and $M(a_1a_2)$ are codewords which represent the length and 'colour' of the runs a_0a_1 and a_1a_2 respectively and are taken from the appropriate white or black run-length code tables.

The coding procedure is shown in the flow diagram, Figure 8.8. The coding procedure identifies the coding mode that is to be used to code each changing element along the coding line. When one of the three coding modes has been identified according to Step 1 or Step 2, an appropriate codeword is selected from the code table given in Table 8.1.

A further feature of Group 4 is the inclusion of an extension mode which allows a portion of the page to be coded in an uncompressed format. There is no mandatory requirement to switch into uncompressed format if the compression algorithm is actually producing an expansion of the data rather than a compression. The extension mode could be used to introduce other types of coding but so far no others have been standardized.

Only pel transmission density is defined, as the resolution capabilities of the actual mechanisms used to create or to print the image are not restricted to allow the maximum freedom of implementation. It is mandatory to provide 200 lpi (lines per inch) in both dimensions, and optional to provide 240, 300 or 400 lpi. Terminals supporting optional pel transmission densities are required to provide a pel density conversion algorithm which results in low impairment to the image. The basic image format is ISO A4 in vertical orientation and a reproducible area which allows for both European and American paper sizes. (European = 210 mm width by 297 mm length, American = 215.9 mm width by 279.4 mm length.) The guaranteed reproducible area is the area of the original image that is reproduced at the receiver and is not lost due to mechanical tolerances, mainly at the printing station, including the tolerance in the scanning and printing resolution of ± 1%. This area is the same for both Group 3 and

Table 8.1
Code table.

Mode	Elements to be coded		Notation	Codeword
Pass	b_1, b_2		P	0001
Horizontal	a_0a_1, a_1a_2		H	001 + M(a_0a_1) + M(a_1a_2)
Vertical	a_1 just under b_1	$a_1b_1 = 0$	V(0)	1
	a_1 to the right of b_1	$a_1b_1 = 1$	$V_R(1)$	011
		$a_1b_1 = 2$	$V_R(2)$	000011
		$a_1b_1 = 3$	$V_R(3)$	0000011
	a_1 to the left of b_1	$a_1b_1 = 1$	$V_L(1)$	010
		$a_1b_1 = 2$	$V_L(2)$	000010
		$a_1b_1 = 3$	$V_L(3)$	0000010
Extension				0000001xxx

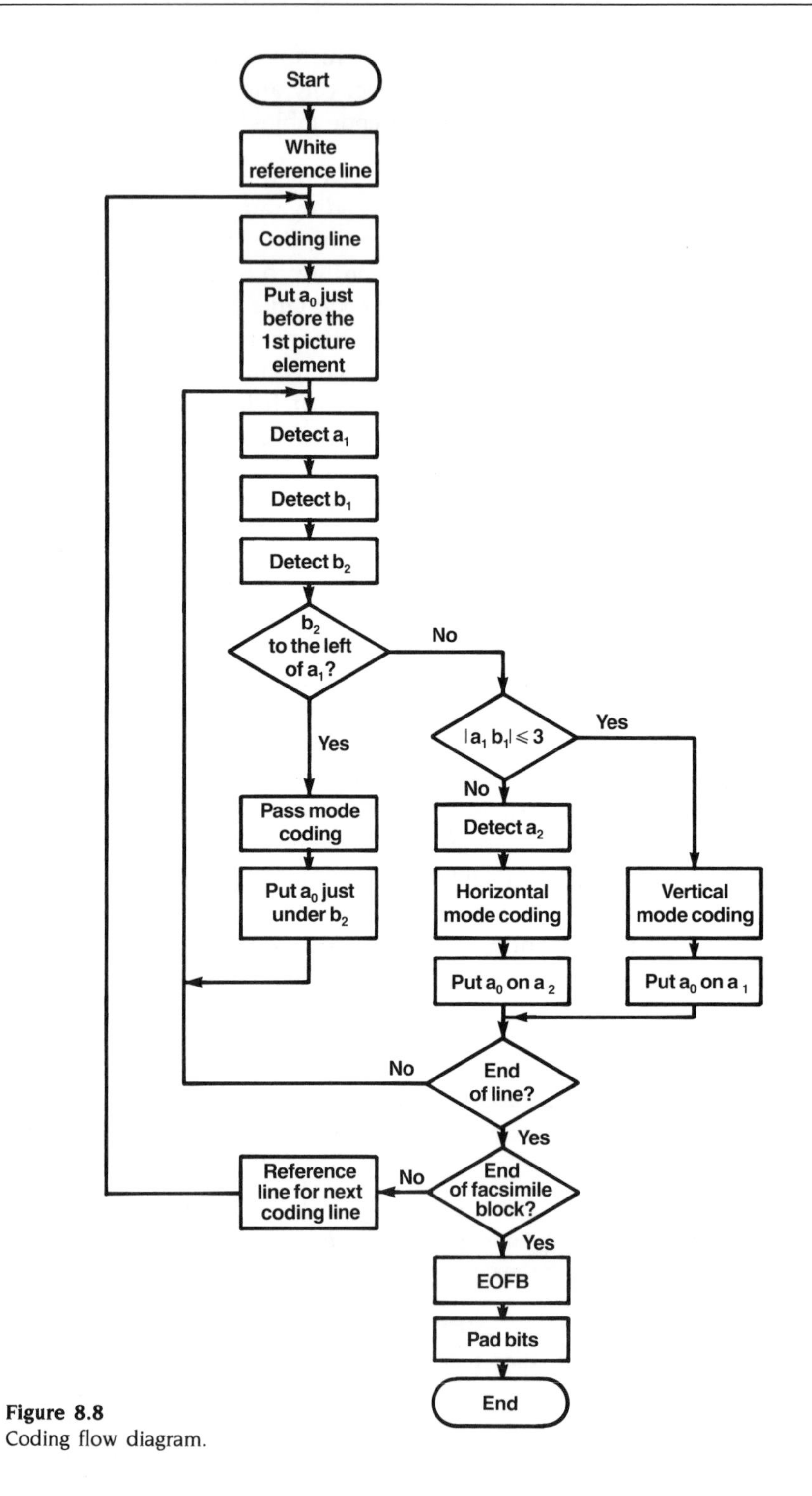

Figure 8.8
Coding flow diagram.

Group 4 facsimile. In the case of European paper it is 196.6 width × 281.46 length in mm.

This close approximation to a full European page is achieved by overscanning the original by 37 pels on both edges, at 200 lpi, and by treating the input paper as being of unknown length until the end of the page is detected, which is usually at the end of the scanning process. The full overscanned image of 1728 pels width is transmitted to the receiver. The unlimited length of page assumption is not a problem for a receiver equipped with a roll-fed printer but for a cut-sheet printer there may be a need to print the image on two sheets occasionally. When an American size image is being received, 215.9 mm width by 279.4 mm long, there is no problem with the length and the full overscanned image width is available at the receiver but the worst case printable area is still the same, therefore there may be some loss of edge detail. However, as the image is centre-line referenced this small loss is not usually a problem and machines are available which make a feature of being able to print the full image width using wider paper, for example.

A terminal identification (TID) is required which enables each terminal to be uniquely identified. This TID is exchanged between the terminals at the start of every connection. It may be used to form the call-identification-line which may be printed at the top or bottom of the first or every page of the transmitted document. Group 4 facsimile terminals are required to support automatic answering.

To handle terminal-to-terminal interactions, Group 4 facsimile provides powerful data communication. The link layer is based upon HDLC LAPB with a few modifications for the circuit-switched type of connection. The network layer is X.25 plus enhancements to bring it in line with the OSI's connection mode network layer definition as far as the ISDN is able to support it. The transport layer is X.224 class 0 and the Session layer is X.225. Above these layers there is the DTAM, Document Transfer and Manipulation, protocol and above that the Open Document Architecture (ODA) which defines the details of the document. All these protocols have a section in their definitions describing how to maintain backward compatibility with the 1984 version of Group 4 class 1 facsimile. As long as the link layer parameters allow over half a second of data to be outstanding and awaiting acknowledgement, then Group 4 works quite efficiently via satellite channels because it generates such large amounts of data.

8.3.3. Facsimile terminal apparatus

Group 4 terminals are not limited to 64 kbit/s or to the ISDN network by the Recommendation but most commercial offerings to date have operated at 64 kbit/s. They have been equipped with either a circuit-switched public data networks interface and a PSTN interface or just an ISDN interface.

Given the high resolutions available in Group 4 facsimile and a high speed network such as the ISDN to transfer the large amount of data created by such resolutions, it would seem logical to employ high quality printing mechanisms in the terminal, e.g. a pre-cut sheet, plain paper, laser-xerographic type. Many terminals have already appeared with this type of printer and the quality of the output image at 400 lpi is for all practical purposes as good as the 300 lpi laser printed original. Unfortunately, there are some disadvantages with this type of printer:

(a) It can only print a full page and not a string of partial pages. Therefore the terminal has to wait until the whole page has been received and decoded before it can transfer that image information to the printing engine and commence printing. This results in a relatively slow printing process of approximately 10 to 15 seconds per page which is between a half and a third of the speed needed to keep up with the 64 kbit/s transmission speed.

(b) If a 400 lpi laser printer is adopted then it is common practice (for compatibility with Group 3) to support only 200 and 100 lpi resolutions as these images are printed by the simple expedient of doubling or quadrupling the ratio of the number of print lines to scan lines. However, this also means that it takes just as long to print a high resolution page as a medium or standard resolution page.

(c) Compared with thermal print engines laser printers are rather expensive. Some manufacturers have retained the termal printing techniques of their Group 3 equipment but with the higher resolutions of Group 4 and thus produced smaller, cheaper Group 4 terminals. However, the quality and stability of the thermally printed image is generally poorer. The limitation of a slow printer can be avoided by installing a large memory to receive the image at maximum speed and then carry out the printing off-line. Some Group 4 terminals have been launched with Winchester disc storage for this purpose and to support mailbox facilities within the terminal.

8.4 PHOTOGRAPHIC VIDEOTEX

Graham Hudson

Basic videotex services, providing, via a telecom network, text and graphics information from a central database on a low cost terminal, have become well established, particularly in Europe. Videotex was invented by British Telecom Research Laboratories in the early 1970s and the first public service, Prestel, was launched in 1979.

The deployment of videotex has been helped by the co-operation between

PTTs in the European Standards body, the Conference of European Postal and Telecommunication Administrations (CEPT). A very comprehensive alpha-mosaic display Standard was produced in 1983. This Standard also contained an option for photographic quality pictures.

8.4.1 Coding

Alpha-mosaic text and graphics displays are character coded; that is each character rectangle on the display is represented by a single code (7 or 8 bits). The information displayed in the character rectangle is produced with reference to a look-up table or a character generator. A full display screen of 24 rows of 40 characters requires 960 bytes which can be transmitted in 6.4 seconds on the PSTN, but on the ISDN this can be achieved in less than a second on the D channel or less than one tenth of a second on the B channel.

With photographic displays each dot (or pixel) has to be independently defined. Facsimile coding is a simple example of photographic coding where each dot can only have one or two values, black or white. With colour still picture photographic coding each pixel can be any colour from a very large range, typically 16 million colours. Each pixel is represented by values of three primary colour components, typically having an accuracy of 8 bits per component. A studio quality picture on a videotex screen has of the order of 262 k pixels and so would require 786 kbytes to be fully represented (see Figure 8.9).

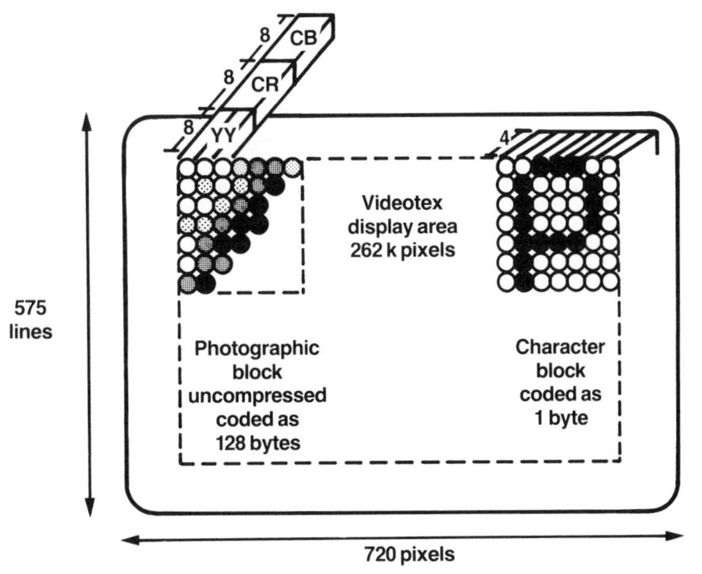

Figure 8.9
Character and photographic coding.

Figure 8.10
Pictures produced by DCT at 0.08, 0.25, 0.75 and 2.25 bits per pel. (Note: originals were in full colour.)

The availability of technologies such as optical storage, graphic display controllers and high speed networks, such as the ISDN, have made it possible to contemplate the storage and tranmission of photographic pictures. Even on the ISDN, however, such a picture would take several minutes to transmit which is unacceptable for interactive services. There has been intense activity in recent years in signal processing research to find the best data compression technique for standardization to make such services possible.

At the end of 1986 experts from ISO and CCITT met to form the Joint Photographic Experts Group (JPEG). They set themselves the task of selecting a high performance universal compression technique. Applications that were kept in mind during the studies, as well as photographic videotex and photographic teletext (broadcast systems), were still pictures for teleconferencing, slow scan pictures for security, medical images, newspaper pictures and satellite weather and surveillance pictures.

A technique was chosen by specifying the main functionality required, setting a number of tests and then evaluating the competing techniques. The selection process took place betwen 1987 and 1988. Twelve proposals were registered covering most of the established compression techniques from predictive coding, block coding, discrete cosine transform, vector quantization and combinations of these techniques. Techniques were evaluated on fucntionality, complexity, but most importantly on subjective quality at fixed compression values.

8.4.2 Discrete Cosine Transform (DCT)

The DCT technique produced the best subjective results. By performing certain mathematical transforms on the spatial pixel values in a block of the image it is possible to produce a set of less correlated coefficients in the transform domain. Let us first look at an intuitive approach to the technique. Consider 64 blocks each of 8 × 8 pixels as shown below: consider each white pixel in each block to have a positive value, and each black pixel to have a negative value. Let us then take an 8 × 8 block of pixels from the picture to be encoded. We will then take each of the values in the picture block and multiply it by the corresponding negative or positive value of in one of the blocks below. We will then sum the resulting 64 products and put that value in the place of the block below. Repeating this for each of the blocks below will result in a matrix of 64 elements; the magnitude of each element will indicate how 'similar' the original picture block was to the corresponding block below. Note that as you move to the right the blocks represent increasing horizontal detail or frequency, and as you move downwards they represent increasing vertical detail or frequency.

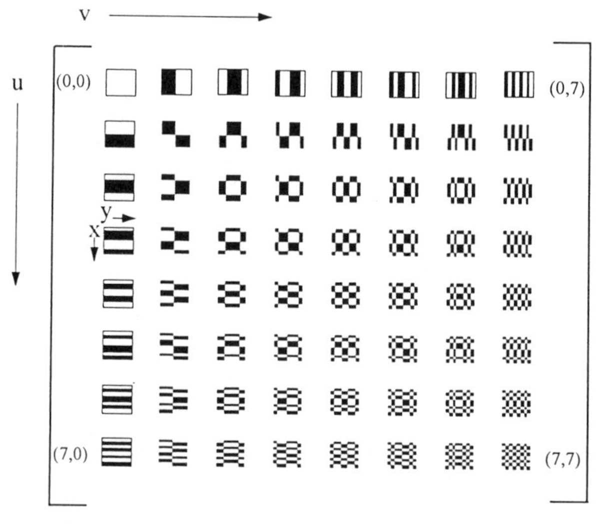

This is a graphical representation of the two-dimensional transform defined by the equation:

$$F(u,v) = \tfrac{1}{4}\, C(u)C(v) \sum_{x=0}^{7} \sum_{y=0}^{7} p(x,y) \left[\cos \frac{(2x+1)u\pi}{16}\right] \left[\cos \frac{(2y+1)v\pi}{16}\right]$$

where $C(u)$, $C(v) = 1/\sqrt{2}$ for u, $v = 0$ and $C(u)$, $C(v) = 1$ otherwise; the two terms in the square brackets are the forward transformation kernel which is shown graphically above and p is the pixel data from the original picture block.

This may be preented in matrix form as: F = [T] [P] [T], where [T]

$$
= \begin{bmatrix}
0.3536 & 0.3536 & 0.3536 & 0.3536 & 0.3536 & 0.3536 & 0.3536 & 0.3536 \\
0.4904 & 0.4157 & 0.2778 & 0.0975 & -0.0975 & -0.2778 & -0.4157 & -0.4904 \\
0.4616 & 0.1913 & -0.1913 & -0.4619 & -0.4619 & -0.1913 & 0.1913 & 0.4619 \\
0.4517 & -0.0975 & -0.4904 & -0.2778 & 0.2778 & 0.4904 & 0.0975 & -0.4157 \\
0.3536 & -0.3536 & -0.3536 & 0.3536 & 0.3536 & -0.3536 & -0.3536 & 0.3536 \\
0.2778 & -0.4904 & -0.0975 & 0.4157 & -0.4157 & -0.0975 & 0.4904 & -0.2778 \\
0.1913 & -0.4619 & -0.4619 & -0.1913 & -0.1913 & 0.4619 & -0.4619 & 0.1913 \\
0.0975 & -0.2778 & -0.4157 & -0.4904 & 0.4904 & -0.4157 & 0.2778 & -0.0975
\end{bmatrix}
$$

In the spatial domain the image energy (square of the pixel values) is normally evenly distributed over the pixel block. Following the transform the majority of the energy resides in a minority of the coefficients. There is much less correlation between coefficients in the transform domain than there was between pixels in the spatial domain. The coefficient $f(0,0)$ the top left-hand corner term, is dominant and contains most of the energy from the pixel block. This zero frequency or d.c. term represents a scaled mean value of the pixel data. The high order coefficients, moving towards the bottom right-hand corner generally contain less energy. Data compression is achieved by reducing the coding accuracy of the less significant coefficient and possibly ignoring some altogether.

The coding process first involves splitting the image into 8 × 8 blocks and then performing the two dimensional DCT on each block (Figure 8.11).

In the ISO standard all coefficients are linearly quantized. Quantization is achieved with reference to an 8 × 8 quantization matrix Q:

Increasing horizontal frequency →

Increasing vertical frequency ↓

$$
\begin{bmatrix}
16, & 11, & 10, & 16, & 24, & 40, & 51, & 61, \\
12, & 12, & 14, & 19, & 26, & 58, & 60, & 55, \\
14, & 13, & 16, & 24, & 40, & 57, & 69, & 56, \\
14, & 17, & 22, & 29, & 51, & 87, & 80, & 62, \\
18, & 22, & 37, & 56, & 68, & 109, & 103, & 77, \\
24, & 35, & 55, & 64, & 81, & 104, & 113, & 92, \\
49, & 64, & 78, & 87, & 103, & 121, & 120, & 101, \\
72, & 92, & 95, & 98, & 112, & 100, & 103, & 99,
\end{bmatrix}
$$

Coefficient quantization matrix

Each value $Q(u,v)$ represents the quantizer step size for that position of

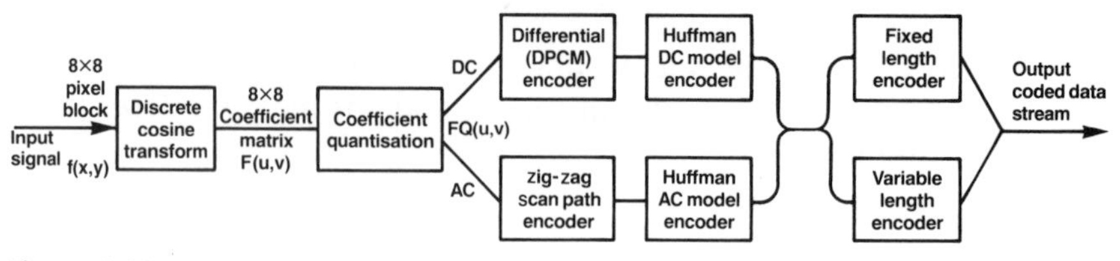

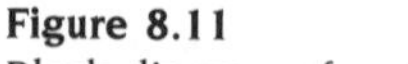
Figure 8.11
Block diagram of cosine transform encoder.

coefficient. The step size has been derived according to the perceptual threshold of the visual contribution of the cosine basis function. A different quantization matrix may be used for each colour component.

The d.c. coefficient $f(0,0)$ of the block, because of its high correlation with adjacent blocks, is differentially encoded with respect to the d.c. term in the previous block. The 63 a.c. coefficients in the matrix are ordered into a one-dimensional sequence according to a zig-zag scan.

Increasing horizontal frequency →

Increasing vertical frequency ↓

$$\begin{bmatrix} (DC) & 1 & 5 & 6 & 14 & 15 & 27 & 28 \\ 2 & 4 & 7 & 13 & 16 & 26 & 29 & 42 \\ 3 & 8 & 12 & 17 & 25 & 30 & 41 & 43 \\ 9 & 11 & 18 & 24 & 31 & 40 & 44 & 53 \\ 10 & 19 & 23 & 32 & 39 & 45 & 52 & 54 \\ 20 & 22 & 33 & 38 & 46 & 51 & 55 & 60 \\ 21 & 34 & 37 & 47 & 50 & 56 & 59 & 61 \\ 35 & 36 & 48 & 49 & 57 & 58 & 62 & 63 \end{bmatrix}.$$

Zig-zag scan path matrix

Both a.c. and d.c. terms are Huffman variable length encoded using a two-stage process—modelling and coding. Each non-zero a.c. coefficient is modelled in combination with the run length of the zero value a.c. coefficients proceeding in the zig-zag sequence. Each run length/non-zero coefficient combination is represented by two codes (see Figure 8.12).

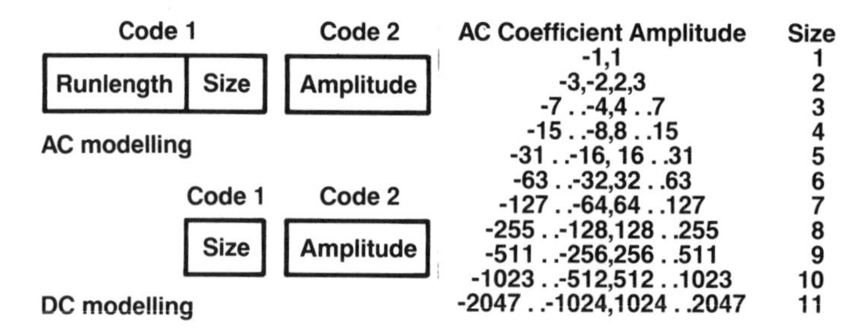

Figure 8.12
Code structure and Huffman code table.

8.4.3 Functionality of the data compression technique

The main requirements of an image date compression technique are that it should be efficient (high compression), economic (low cost of implementation), fast (capable of real transmission time decoding) and be applicable to a broad range of natural monochrome and colour picture services.

It was felt that the maximum acceptable waiting time for a picture to be transmitted was 5 seconds. For videotex purposes with a display area of the order of 512 × 512 pixels a compression of the order of 1 bit/pixel is required. The ISO DCT provides good quality pictures with a compression of 0.5–1 bit/pixel (Figure 8.13).

Most traditional coding schemes, particularly those delivering to output devices such as printers, produce a sequential picture build-up. That is the picture is constructed in full resolution and colour pixel by pixel from left to right and top to bottom of the image. Such a scheme is ideal for file transfer between computers and for applications such as ISDN photovideotex where the picture build-up on the screen is very fast (Figure 8.14).

For very high resolution pictures on the ISDN and for transmission over slower data rate networks a progressive coding scheme is very desirable. The method provides a crude picture quickly which can, if required, be subsequently improved in several stages until the final quality is achieved. A picture of increasing resolution and accuracy is provided which is useful in database applications where it can match the amount of information transmitted to the capabilities of the output device. This also allows for easy scanning/browsing of databases by only viewing either a low quality full size picture or a reduced size (sub-sampled) picture.

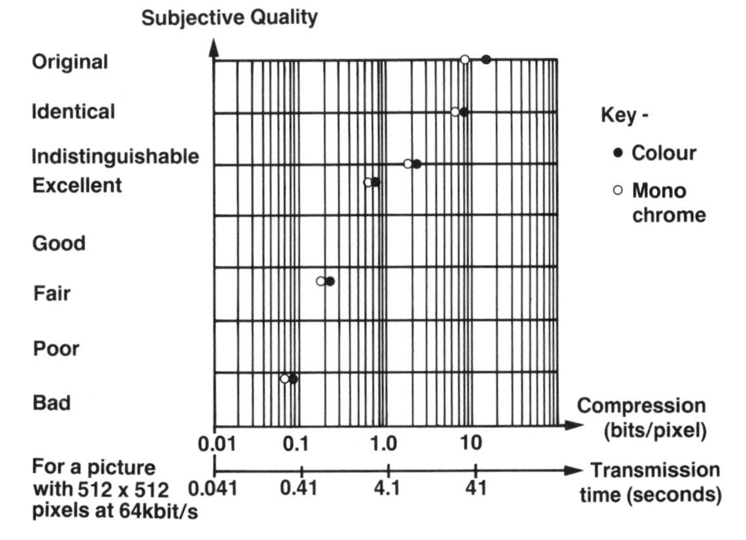

Figure 8.13
DCT picture quality for different compression values.

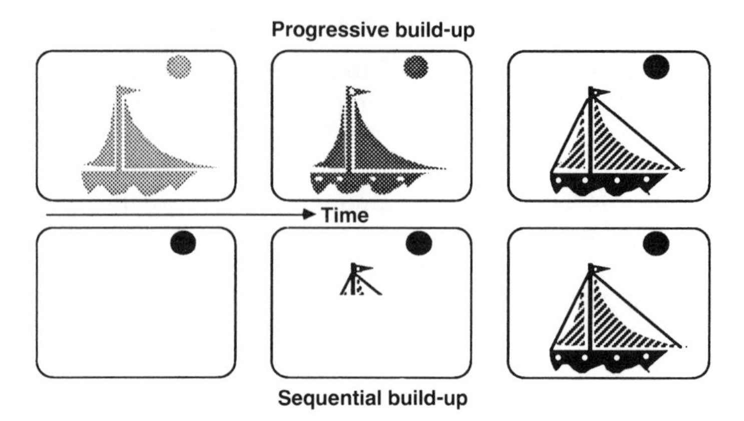

Figure 8.14
Progressive and sequential picture build-up.

Easily recognizable pictures can be achieved with a compression of less than 0.1 bit/pixel. This compresses an image to 3.3 kbytes and can be transmitted on the B channel in under half a second and in 3 seconds on the D channel.

Pictures that are subjectively indistinguishable can be obtained with a compression of the order of 2 bit/pixel (64 kbytes, 8 seconds on a B channel). Some applications such as the transmission of medical images and surveillance pictures require the final image to be identical to the original. Typically this can be achieved with a compression of the order of 8 bit/pixel.

Not so long ago it was thought that the complexity/cost of implementation of a high performance compression technique such as a cosine transform would be prohibitive for high speed real time transmission schemes. High speed microprocessors (16 bit, >20 MHz), digital signal processors or VLSI matrix multipliers can each be used to decode DCT at 64 kbit/s.

8.4.4 ISO DCT conformance levels

The DCT compression technique is encapsulated in an ISO Standard and a CCITT Recommendation. The ISO Standard has three conformance levels —a baseline system, an optional extended system and an independent function level.

The baseline system must be implemented by every ISO encoder and decoder. It also serves as the default mode and guarantees image communication between any standard encoder/decoder. The system is limited to sequential picture build-up for 8 bit/pixel and does not allow for identical (lossless) coding. Default Huffman tables and quantization matrices based on the CCIR 601 colour space (Y, C_R, C_B) are specified. For a description of this colour space see Section 8.5. There are two default quantization matrices, one for the Y component and the other for the C_R and C_B components. There are four Huffman tables—$Y_{d.c.}$, $Y_{a.c.}$, $C_R/C_{Bd.c.}$ and $C_R/C_{Ba.c.}$.

The extended system includes all the baseline system features but allows for a number of significant enhancements. Improved coding efficiency can be provided by using arithmetic coding or non-default Huffman coding. A variety of progressive build-up features are allowed for including hierarchical and non-hierarchical coding, bit slicing (increasing accuracy) and spectrum selection (increasing resolution). Reversibility is provided by sending a difference signal between the non-exact DCT image and the original image.

The independent function allows for a stand alone method with reversible (lossless) compression for applications requiring only this capability and uses a simple two-pixel predictor.

8.5 VIDEO

Geoff Morrison

This section describes how moving video pictures are digitally coded and compressed for ISDN bit rates. The algorithm was standardized by CCITT as Recommendation H.261 and is primarily intended for use at bit-rates between about 40 kbit/s and 2 Mbit/s. For videotelephony and videoconferencing the coded video bit stream would be combined with audio in a multiplex structure conforming to Recommendation H.221 which provides framing and various housekepeing channels. Several options are available for the audio coding algorithm and the bit-rate apportioned to it, so that the balance between audio and video qualities can be adjusted for any given total bit-rate. For example, the use of 16 kbit/s audio permits 46.4 kbit/s of video to be conveyed with it in a single B channel. In the double B case, retaining the same 16 kbit/s audio more than doubles the rate available for video to yield significant improvements in picture quality. Another option is to utilize audio coding according to CCITT Recommendation G.722 at rates of 48, 56 or 64 kbit/s. This algorithm can provide 7 kHz audio bandwidth (see Section 8.1) and would be the preferred choice when the total bit-rate exceeds a few hundred kbit/s. An overall block diagram of the video codec is given in Figure 8.15.

A normal broadcast television picture is built up from 625 or 525 lines repeating at 25 or 30 times per second respectively. These two scanning standards originated in Europe and the USA many years ago and have since been adopted throughout the rest of the world, each country choosing one of the two. A colour television picture can be considered a combination of three signals representing red (R), green (G) and blue (B) constituents. For black and white displays a luminance signal Y can be derived:

$$Y = 0.30R + 0.59G + 0.11B.$$

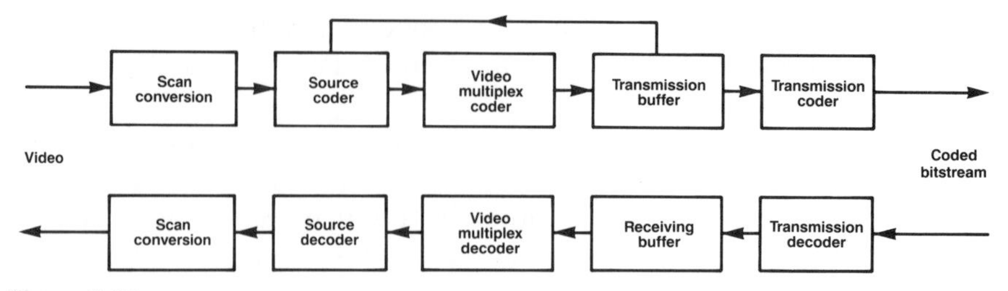

Figure 8.15
Outline block diagram of video codec.

Colour information is then carried in two colour difference signals C_B and C_R:

$$C_B = B - Y$$

and

$$C_R = R - Y.$$

By suitable combinations of Y, C_B and C_R a return can be made to R, G and B for colour displays. The purpose of this somewhat devious process is to take into account the fact that the human eye is rather more sensitive to luminance detail than to colour detail and hence the colour differences can be allocated less bandwidth. A straightforward way, such as specified by the International Radio Consultative Committee in CCIR Recommendation 601, to digitally encode such a picture is to sample Y at 13.5 MHz and each of C_B and C_R at 6.75 MHz, each sample being represented with 8 bits resulting in an overall bit rate of 216 Mbit/s. This exceeds the capacity of a 64 kbit/s ISDN B channel by a factor of 3375. Obviously such massive compression needs concerted effort.

The coding scheme achieves its compression by a combination of several techniques. Some, for example entropy coding with variable length codes, achieve data reduction in a lossless way; that is, the coding is completely reversible and achieves an exact reproduction of the original. The others are lossy compression methods, such as low-pass filtering, which are not reversible and produce only an approximation to the original input. These always give an objective loss, but subjectively the distortion can range from being invisible to highly annoying and part of the science of video codec design is in the exploitation of the properties of the human visual system to minimize the subjective loss of quality.

8.5.1 Source format

The first technique used is the simple one of sacrificing a little of the picture sharpness. In the horizontal direction, the luminance is sampled at 6.75 MHz, half of CCIR 601, allowing a bandwidth of just over 3 MHz to be obtained.

For vertical resolution, the CCITT sought a recommendation for world-wide use and therefore needed to take account of the 625 and 525 formats. Rather than have two different versions of the recommendation the decision was taken to specify just one based on a compromise in which the number of lines bears a simple relationship to the 625 system and the repetition rate comes from the 525 system. The adopted number of luminance lines is 288, being half the number of the active lines in 625. In that system some 575 lines are avilable for the picture itself, the others giving a retrace interval to allow the scanning spot to move from the bottom back to the top of the picture. For 525 sources and displays the conversion is a little more complex, but the ratio 5:3 allows 480 lines to be covered. Theoretically this is just a few lines less than the active height of a 525 line picture but in practice displays are always overscanned and the missing lines would not be seen anyway. The picture refresh rate adopted comes directly from the 525 system, being approximately 29.97 Hz.

For the colour differences a further reduction of spatial resolution in each direction to half of the luminance is employed, taking advantage of the eyes lower acuity to colour detail mentioned above. The objective spatial resolution of the coding format is hence much less than CCIR Recommendation 601. However, the subjective loss is not as drastic as the numbers might suggest. The horizontal bandwidth of more than 3 MHz is on a par with the resolution of the colour display tubes typically found in domestic television sets and is better than many home video cassette recorders. Also, in the vertical direction resolution is better than the 1/2 of 625 or 3/5 of 525 could imply because those two formats use interlace whereas H.261 does not. The explanation of this is outside the scope of this book.

CCIR Recommendation 601 specifies a digital active line of 720 pels of luminance and 360 of each of the two colour difference signals. Simple 2:1 reductions would give 360 and 180. However, for reasons which will become obvious below it is desirable to have numbers which are an integer multiple of 16 for luminance and 8 for colour differences. To achieve this some pels are discarded at the edges of the picture to leave 352 and 176. The loss of picture information is minimal as the ignored pels frequently fall outside the real picture area in the part allocated to horizontal flyback of the scanning spot and even if not would not be visible on normally overscanned displays.

The format came to be known as 'Common Intermediate Format' and referred to as CIF. Subsequently, to permit simpler codecs for use with smaller screens and low bit-rates, for example ISDN videotelephones, an additional format was introduced whose spatial resolution is one half of CIF in each axis. This factor of four prompted the term 'Quarter CIF' (QCIF).

PCM encoding of CIF with 8 bits per pel would need $352 \times 288 \times (1 + 1/4 + 1/4) \times 29.97 \times 8 = 36.5$ Mbit/s.

Before leaving the source coding format, it should be noted that H.261 does not specify how CIF or QCIF is derived from the coder input video standard nor how a decoder should convert the decoded images in CIF or QCIF form to a suitable format for displays.

8.5.2 Source coding

A block diagram of the source coder and decoder is given in Figure 8.16. The function of the coder is to achieve data compression by redundancy reduction. The switches are shown in the position they occupy most of the time to take advantage of the fact that the television signal is a series of snapshots and that each one is very much like the one before it. Indeed for stationary parts of the scene the difference between successive pictures is zero and no information needs to be transmitted at all. The block marked P is a memory which holds the previous picture. By making the delay through it plus the other functions in the loop from the output of the subtractor back to its lower input exactly equal to the period of one picture, the corresponding points in successive pictures are compared. Only in places where there have been changes such as motion will there be a non-zero output which needs to be transmitted.

A further refinement is motion compensation. If a moving object is undergoing purely translational motion then that part of the picture can be constructed by reproducing an offset part of the previous picture. The offset will depend on the direction and magnitude of the movement and is called a motion vector because it has two components, left–right and up–down. The vector is used to adjust the delay of P so that corresponding

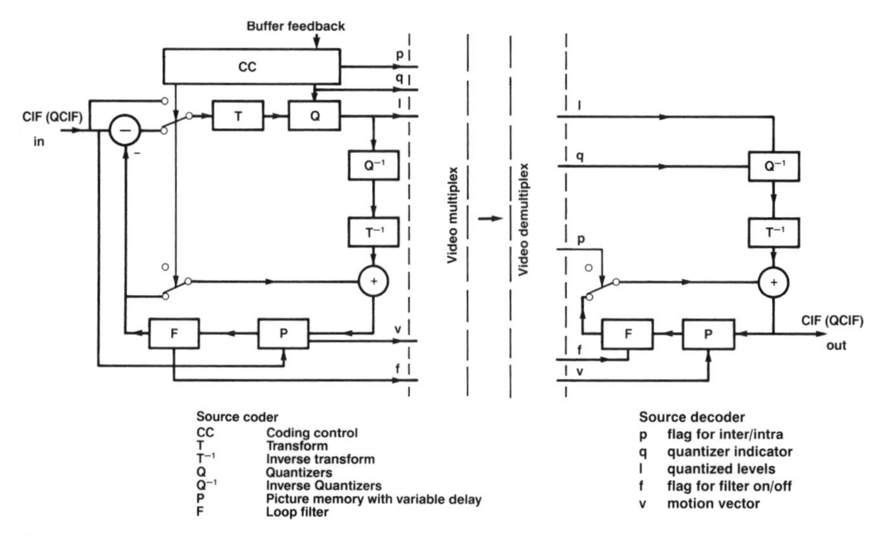

Figure 8.16
Source coder and decoder.

points on a moving object are fed to the subtractor. Transmitting a motion vector for individual pels would require too many bits so the picture is divided into blocks of 16 × 16 luminance pels and one vector is used for it and also for the two spatially corresponding 8 × 8 colour difference blocks.

This process is known as motion compensated inter-picture prediction. The prediction is based on the previous picture and the whole process is recursive. The subtractor output, the prediction error, will not always be zero. Moving objects can change shape or rotate and background can be uncovered at the object edges so that one single vector cannot apply perfectly for all the block. However, the energy of the prediction error is much less than the original signal and hence more easily transmitted. The process is completely reversible. By accumulating all the error signals and using motion vectors a decoder could reconstruct the original signal exactly. However, to get even more compression it is necessary to apply lossy compression to the prediction error in the quantizer Q. Because the predictions in the encoder and decoder must be the same and the decoder only has decoded pictures to predict from, it is necessary for the encoder to use the same modified versions for its predictions. This is the reason for the decoding boxes Q^{-1} and T^{-1} and the adder being required in the coder.

The method of determining motion vectors at the coder is not defined. The problem is essentially one of pattern matching and can be very expensive computationally since the vectors are allowed to have components in the range − 15 to + 15. This means that for each block up to 225 comparisons of 16 × 16 blocks could be needed to find the one with the smallest prediction error. Some schemes use simplified methods such as tree searches.

Because the coding scheme is recursive there are the problems of start-up and transmission errors. Under these conditions the loops at encoder and decoder do not contain the same information and as the decoder is essentially a perfect integrator the differences would remain forever. To deal with this the switches are occasionally flipped over so that the video signal itself instead of a difference is sent. However, this intra-coding mode requires many more bits and so can only be invoked for a small fraction of the time.

The block marked T performs a Discrete Cosine Transform (DCT) on 8 × 8 blocks according to the formula as set out in Section 8.4.2.

This reversible mathematical process converts the 64 numbers representing the levels of 64 pels to another set of 64 numbers, transform coefficients, but the information is contained in them in a different form such that coding gain can usually be achieved. These transform coefficients represent the amplitudes of spatial frequencies in the block. For example, $F(0,0)$ is the average level of the block and $F(1,0)$ give a measure of the difference between the left and right halves. Many blocks contain only a few frequency components or several of them have low enough amplitudes to be ignored such that it is not necessary to transmit all 64 of them.

The quantizer, Q in Figure 8.16, introduces lossy compression by forcing the transform coefficient amplitudes to adopt a restricted set of values which require fewer bits to differentiate between them. Of course this introduces an error in the coefficient but the inverse transform spreads it over all the pels in a block so the effect is much less than if the pels themselves had been subjected to the quantization process. Q actually contains 31 different quantization laws. All are linear but have different step sizes so that more or less effect can be obtained. This facility is most useful to allow the intrinsic non-constant bit-rate of the coder to be matched to the fixed bit-rate transmission channel as will be mentioned later.

Since the forward transform is only in the encoder it does not have to be precisely specified. However, an inverse transform (IDCT) is present in both the encoder and decoder and if the two do not give the same outputs then the differences will accumulate and the loops diverge with visible effects such as a stationary granular noise pattern on the pictures at the decoder. Unfortunately the IDCT, like the DCT, contains cosine terms and many of them cannot be represented exactly with finite arithmetic. Errors will therefore be introduced by rounding or truncation in real systems. To allow various implementations of the IDCT function to be used, the CCITT decided not to specify a unique fixed precision IDCT but to leave a small error tolerance. Long-term accumulation of the allowed small mismatch errors between different IDCTs at encoder and decoder is controlled by stipulating that a block must be intra-coded at least once in every 132 times it is transmitted.

The filter, F in the diagram, is a two-dimensional slow roll-off low-pass filter which can be used to suppress noise introduced by the quantizer to prevent it circulating in the loop. However, it cannot be left active in all blocks because stationary detail would be affected, giving a poor prediction and an increase in the number of transmitted blocks with a consequent increase in bit consumption and therefore poorer rendition of the genuinely changed areas. In many coders, therefore, the filter is linked to the motion vector so that it is switched to all-pass when the vector is zero.

The final item in Figure 8.16 is coding control. This determines the coding parameters such as the selection of the quantizer and the use of intra-mode. Additionally, especially at low channel rates the scheme is not able to transmit the input picture rate of 30 Hz and picture dropping is invoked. Since the decoder only needs to know the decisions made, but not the reasons for them, the coding control strategy is not specified.

8.5.3 Video multiplex coding

The function of the video multiplex is to serialize the various data streams such as those containing addresses of changed areas of pictures, motion

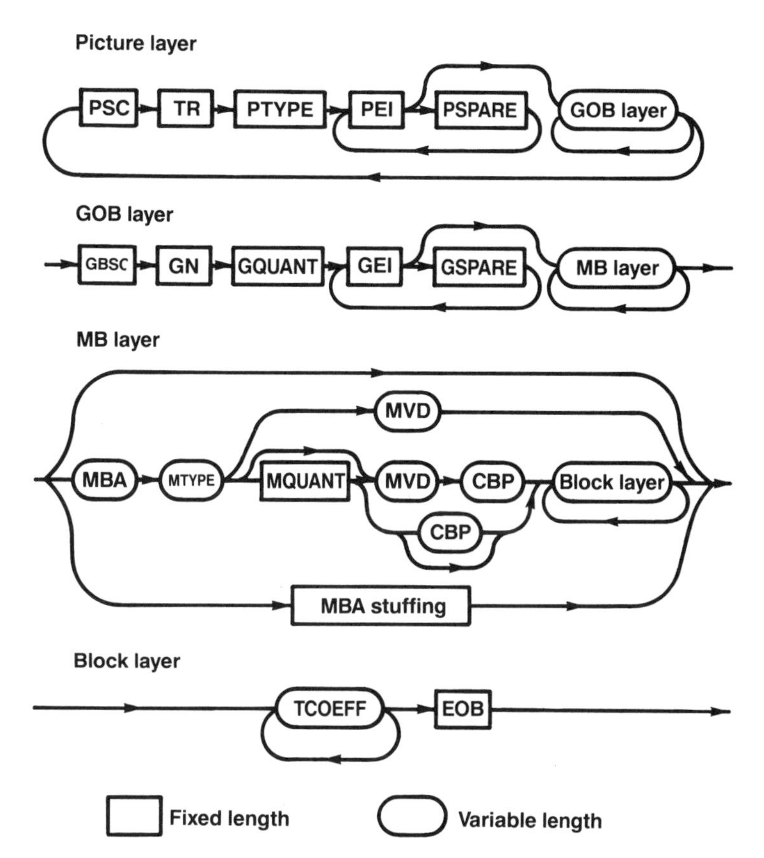

Figure 8.17
Syntax diagram for the video multiplex coder.

vectors, quantized coefficients and so on into one bit stream. It also achieves further lossless data compression by statistical coding of the data from the source coder. The multiplex is arranged in a hierarchical structure with the four layers—Picture, Group of Blocks (GOB), Macroblock (MB), Block—as shown in the syntax diagram of Figure 8.17.

The Picture Start Code (PSC) is a specific 20-bit word which allows the decoder to identify the start point of the picture. Next comes the 5-bit Temporal Reference (TR) which is a form of time stamp to allow the decoder to put pictures to the output screen at the correct times if picture dropping has been invoked by the coder. PTYPE contains 6 bits of information which relate to the complete picture such as whether it is CIF or QCIF format. PEI is a single bit which indicates whether PSPARE of 8 bits and another PEI bit follow. This will permit a simple linked list structure to be used at a future date for enhancements but initially only the first PEI bit is included and indicates that PSPARE is absent. Information for the GOB layer then follows.

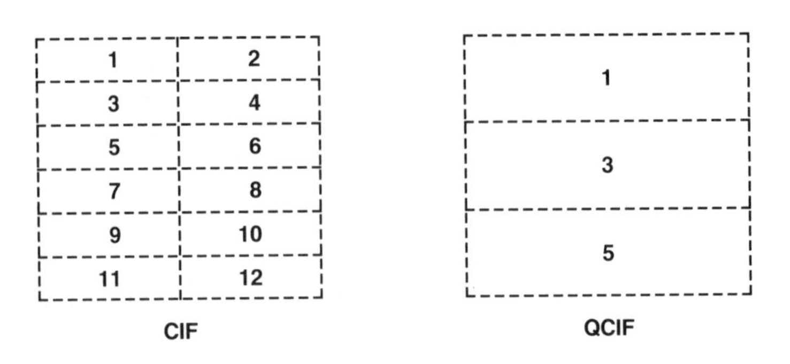

Figure 8.18
Arrangement of GOBs in a picture.

A group of blocks comprises one twelfth of the CIF or one third of the QCIF picture areas as shown in Figure 8.18. A GOB thus covers 176 pels by 48 lines of Y and the spatially corresponding 88 by 24 of each of C_B and C_R. The Group of Blocks Start Code (GBSC) is a 16-bit word and is immediately followed by 4 bits containing the GOB number (GN) to identify which part of the picture the GOB is from. GQUANT is 5 bits to signal which of the 31 quantizers is in use in the GOB. GEI and GSPARE allow a linked list extension structure in the same manner as in the picture layer. Then comes the macroblock data.

Each GOB is divided into 33 macroblocks as shown in Figure 8.19. A macroblock relates to 16 lines by 16 pels of Y (four 8×8 blocks) and the spatially corresponding 8×8 of each of C_B and C_R as in Figure 8.20. Macroblock Address (MBA) is a variable length codeword indicating the position of a macroblock within the GOB. MBA is the differential address between transmitted macroblocks. Macroblocks are not transmitted when that part of the picture is not to be updated. The variable length code table for MBA contains an extra codeword called MBA stuffing. This should be discarded by decoders as its function is merely to pad the bit stream as may be necessary when pictures with little motion are coded at the higher bit-rates.

MTYPE is a variable length codeword to give information about the macroblock and which data elements follow. For example, it indicates whether the macroblock is inter- or intra-coded, whether there is a motion vector or not and whether the filter is applied in the prediction or not. It can also signal that a different quantizer will be brought into effect from this macroblock onwards. For a macroblock which can be perfectly reconstructed just by moving part of the previous picture only Motion Vector Data (MVD) is necessary. When the motion vector is zero this is explicitly signalled by MTYPE and MVD is not included. MVD is not needed for intra-coded macroblocks since these do not use predictive coding. For data compression reasons MVD is a variable length codeword representing the

1	2	3	4	5	6	7	8	9	10	11
12	13	14	15	16	17	18	19	20	21	22
23	24	25	26	27	28	29	30	31	32	33

Figure 8.19
Arrangement of macroblocks in a GOB.

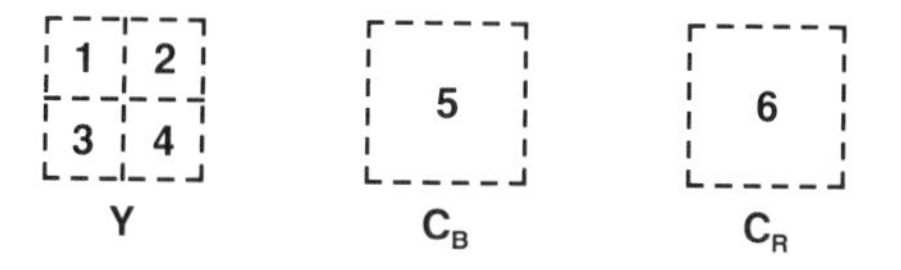

Figure 8.20
Arrangement of blocks in a macroblock.

vector difference from the neighbouring macroblock since there is a high probability that its motion vector will be similar.

Coded Block Pattern (CBP) is a variable length codeword which tells which of the six 8×8 transform blocks have coefficient data sent for them in the block layer. CBP is not needed by all macroblock types. For example it is always the case that intra-coded ones have coefficients for the six blocks.

Data for a block consists of codewords for quantized transform coefficients (TCOEFF) in the zig-zag order given in Section 8.4.2, followed by an End of Block (EOB) marker. The marker is needed as the number of coefficients may be between 1 and 64 because the trailing zero amplitude ones are not sent. The zig-zag follows the path of increasing probability that a coefficient is zero. A two-dimensional variable length coding method is adopted in which the 64 most commonly occurring combinations of successive zeros (RUN) and the following level (LEVEL) are encoded with words of between 2 and 14 bits. The remaining combinations of (RUN, LEVEL) are encoded with 20 bits consisting of a 6-bit ESCAPE, 6-bit RUN and 8-bit LEVEL.

The nature of the coding algorithm means that the rate of generation of bits is not constant but varies depending on the input pictures. Pictures with more motion have more changed areas and the statistics of these changes will be reflected in the lengths of the variable length codewords transmitted. Furthermore, the motion is rarely distributed evenly over the picture area so that the generation is spasmodic. A buffering process is necessary at the encoder to match this varying rate to the fixed channel rate and a corresponding inverse process at the decoder. Such buffers introduce delays and for conversational services such as videotelephone and videoconference these must be bounded at a few hundred milliseconds. The buffer, therefore, can only perform short-term smoothing and the long-term average data rate must be controlled by altering parameters in the coding loop. As mentioned earlier coarse control can be achieved by dropping complete pictures and fine control by selection of the quantizer, so most coders incorporate a feedback mechanism from the buffer fullness to regulate those functions.

The coded bitstream is rather vulnerable to transmission errors because of its heavy use of variable length codewords. In addition to corrupting one word, a single-bit error can make one transmitted variable length codeword appear to be two, or vice versa. This results in a decoder losing track of the correct data sequencing and interpreting all subsequent data wrongly, usually until the next GOB Start Code. Because of the recursive nature of the algorithm any disturbances remain on the screen for many picture periods. Further the use of motion vectors can make corrupted regions move around the screen and grow larger. To offer protection against this, the transmitted bit stream contains a BCH (511,493) Forward Error Correction Code which permits two random errors or a burst of up to six errors in a 511-bit block to be corrected. To allow the video data and error correction parity information to be identified by a decoder an error corrector framing pattern is included. This consists of a multiframe of 8 frames, each frame comprising 1-bit framing, 1-bit fill indicator, 492 bits of coded data and 18 bits parity as shown in Figure 8.21. The fill indicator can be set to zero to signify that the following 492 bits are not video but merely stuffing bits which the decoder should discard. This can be used in a similar manner to MBA stuffing described above.

The algorithm is intended for use over a 40 to 1 range of bit-rates and the resultant overall quality will depend markedly on the operating rate and on

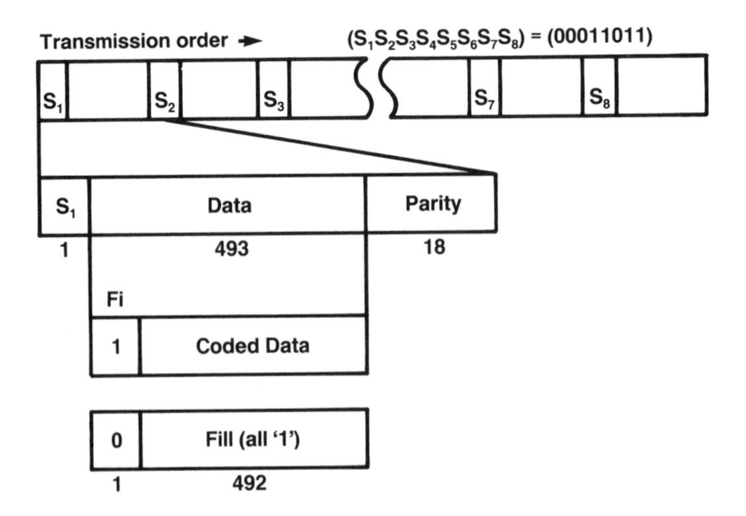

Figure 8.21
Error correcting frame.

the characteristics of the input video signal. In addition, the H.261 Recommendation only specifies those items which are necessary to guarantee compatibility and leaves considerable freedom to coder designers to achieve what they feel to be the optimum balance of the various distortions that must be introduced to achieve a given bit-rate. This means that even when coding the same pictures at the same bit-rate two coders can yield significantly different results. Unfortunately, even with photographs, it is not possible in a book to show examples of the quality that can be obtained because the motion is such an important part of the subjective evaluation. However, at 1.5 or 2 Mbit/s the pictures compare favourably with home video recorders even when coding typical broadcast television programmes. At 64 kbit/s a fixed camera position is really required and pictures can become noticeably jerky due to the necessarily low coded picture rate, but they are suitable for the intended purpose of interactive audiovisual communication.

8.6 AUDIO VISUAL SERVICES AND EMBEDDED DATA

The applications so far described may stand in isolation but in the future the user will often wish to switch from application to application during a call or have a combination of applications sharing the available bandwidth. For example, during a speech call the user may wish to send at a low data rate text or a facsimile picture and having displayed the text or picture the sender may wish to control an electronic pointer to highlight a particular area on the destination display. In audio and video conferences, control signals may be required so that the chairman can control the conference and speakers may wish to share a graphical display or telewriter.

These are examples of multimedia and multipoint operations and what is clearly needed is a control channel in an agreed format which will allow the sender to indicate to the other participant or participants how the available bandwidth is divided and what is being carried in each of the sub-channels. Although the ISDN is primarily based on 64 kbit/s channels, we will see in the next chapter that in some circumstances multiple 64 kbit/s channels may be assembled and it would be convenient if the control channel could also be used to control the partitioning of more than one 64 kbit/s channel.

One way of providing such a control channel is via the signalling D channel and one of the features of that channel is to allow the passing of user-to-user information. However the primary use of the D channel is to convey signalling information controlling the set-up of the calls which inherently demands the use of extensive processing power within the network. The use of these processors simply as store-and-forward packet switches makes poor use of the investment and if used extensively for this purpose could have a serious impact on the dimensioning of exchange processor utilities.

The alternative is to use a signalling protocol within the 64 kbit/s channel and this is usually known by its CCITT Standard number—H.221. This uses the least significant bit of each 8-bit octet, very much in the way that early network signalling in the USA 'stole' bits (see Section 3.1). However, in this case the stealing is at the behest of the user and not the network. By taking this least significant bit of each octet a Service Channel may be derived of 8 kbit/s.

This service channel is split into frames of 80 bits. The first 8 bits of each frame contain a Frame Alignment Signal (FAS), and perform the following functions:

— frame alignment, using an alignment word unlikely to be imitated by information signals;
— optional multiframe alignment (16 frames);
— optional multiframe counter (modulo 16) for synchronization of multiple connections (even if some are via satellite and the others terrestrial);
— connection numbering;
— optional cyclic redundancy code checking (CRC4), for error detection on the whole bit stream;
— the 'A-bit', indicating frame/multiframe/synchronism in the opposite transmissions direction.

In connections of 384, 1536, and 1920 kbit/s the service channel is provided only in the designated 'timeslot 1'.

The next eight bits are known as the bit allocation signal (BAS), which provides three distinct functions:

— capability codes indicating the range of ability of a terminal to decode the various types of information signal: narrow/wideband audio, various video parameters, transmission and data rates, encryption;
— command codes specifying the exact contents of each transmitted frame;
— escape codes providing extension for a variety of purposes.

Clearly the BAS codes must be very error-resilient. To achieve this, they are only transmitted in alternate frames, the intervening frames carrying an 8-bit code capable of correcting two errors. Thus even for a random-error rate of $1:10^3$ the probability of a BAS code error is still extremely small. Burst errors and slip are more problematic, and for this reason the valid codes must be retransmitted from time to time to make sure the receiver is kept in track.

The remaining capacity of the service channel (6.4 kbit/s) and of all the other capacity (56 kbit/s) is used for a variable mixture of audio, video and data. It can be seen that the overhead of the frame structure is low—only 1.6 kbit/s for FAS and BAS in each connection. If encryption is in use, the initialization vectors are transmitted in the eight bits of the service channel following BAS.

In operation, both receivers are continuously searching for frame alignment, and when this is achieved the A-bit is set to zero on the outgoing channel. Only when receiving A = O can a terminal be sure that the remote terminal can understand and act upon a BAS code. The procedures for using the BAS codes are set out in Recommendation H.242. The terminals exchange their capabilities and then adopt a transmission mode suitable for the application. Provided that it remains within the capability range indicated by the other terminal, the content of a transmitted signal may be varied at will, using BAS commands.

Each command is effective from the next frame pair; that is, the frame following that in which the error-correction code for the BAS command has been transmitted. Since the frame rate is 100 Hz, a change can be made every 20 mS.

For example, if a terminal has received the capability codes 100 00100 and 101 00100, it knows that the other terminal can decode wideband (G.722) audio at 48, 56 or 64 kbit/s and can also accept a data signal at 64 kbit/s in the service channel. If then it chooses to transmit 56 kbit/s audio with 64 kbit/s data, it must use the appropriate commands 000 11000 and 011 00100.

If both terminals have indicated a capability to receive at higher transfer rates than can be transmitted on the available connection, then additional 64 kbit/s connections may be set up.

A multipoint call can be treated as a number of bidirectional point-to-point calls between the terminals and a special bridging unit known as a 'Multipoint Control Unit' (MCU); a simple MCU transmits to every terminal the mixed audio signals from all other terminals, and broadcasts any data (from one terminal at a time). If there is video, this is switched such that each participant receives one of the other pictures. The H.221 frame structure and the H.242 procedures are used on each connection in a very similar way to point-to-point calls—further details can be found in Recommendations H.231 and 243.

8.7 CUSTOMER PREMISES EQUIPMENT

There is a chicken and egg relationship between network and terminal availability. The lead time for network developments is many years, whereas terminals come and go in a year or two. Thus any review of terminals can only be transitory.

ISDN terminals can be classified into several categories.

Telephony

There is no immediate pressure to adopt ISDN for voice transmission purposes. The reasons for the development of ISDN telephony terminals are fivefold:

1. Without the analogue local loop the quality is marginally improved, and for loudspeaking telephones the absence of a two-wire section means that one source of feedback and instability is eliminated.
2. The improved signalling capability gives easy access to facilities on the public network similar to those found on modern PABXs including displays of the calling line identity. This is particularly relevant for CENTREX operation where the public exchange is used to emulate a PABX.
3. The telephone may incorporate a terminal adapter function offering a range of data ports.
4. Encryption may be much more effectively added to a digital service in comparison to that available through analogue working.
5. To complete the range of equipment offered by a supplier which can be associated with other I.420 products. The irony of this is that as most lines will have a telephone associated with them, the telephone will become the commonest type of terminal equipment even though it gains least advantage.

Terminal adapters, cards and gateways

These allow existing terminals to be connected to the ISDN. The classic adapter is a stand alone unit perhaps making the conversion from a traditional interface, such as RS232 or X.21, to I.420. The adapter may incorporate management functions and a range of interfaces. Figure 8.22 shows such an adapter. Alternatively personal computers may be connected to the ISDN by plug-in cards such as those in Figure 8.23. Gateways connect LANs to other LANs or to individual terminals via the ISDN. A wide range of equipment is available ranging from simple multiplexers which emulate a private circuit up to complex intelligent servers which set up circuits via the ISDN based on the routing and traffic needs of the

Figure 8.22
ISDN terminal adapter for V.24, V.35 and X.21 (photo courtesy IBM).

LANs. Larger versions may use primary rate interfaces rather than basic rate.

The use of ISDN for LAN interconnection challenges the ingenuity of the designer as LANs operate in a connectionless manner and the ISDN is connection orientated. It is not too difficult to derive a strategy for setting up calls when an inter-LAN message arrives, either based on a previously constructed routing table or on some searching process. What is more difficult is to decide when that routing is no longer required as there is no indication from the LAN that traffic to that destination has ceased. In practice some form of time-out is generally used, but setting the time delay too short can lead to frequent additional call set-ups, whilst setting it too long means that calls are held unnecessarily with no traffic. Frame mode services much more nearly match these needs, than do circuit-switched or X.25 based services.

Audio terminals

As discussed in Section 8.1, 64 kbit/s can give fairly high quality audio transmission. Conference units are available which exploit this feature. Broadcasters also make use of this for occasional programme circuits; Figure 8.24 shows Harry Peart of the BBC's World Service reporting the

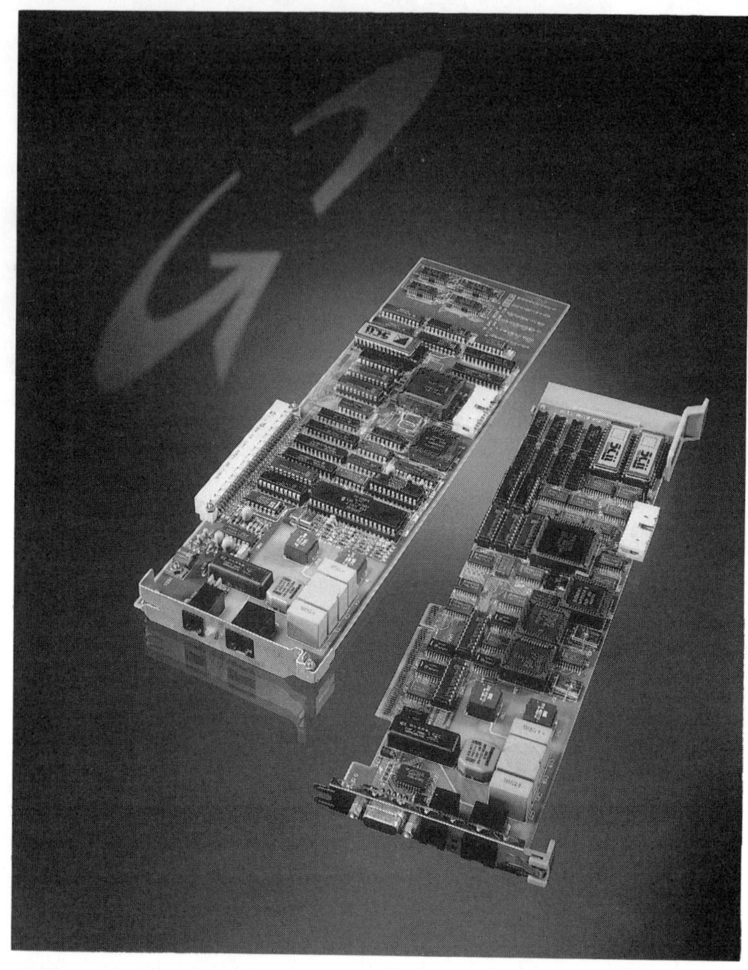

Figure 8.23
ISDN personal computer card (photo courtesy Gravatorn Technology).

1991 British Open Golf Championship at Royal Birkdale. The grey box is an X.21 terminal adapter; the black box is the G.722 speech encoder. As discussed in Section 8.2 the G.722 encoder is designed for speech applications and is also satisfactory for music within its bandwidth limitation of 7.5 kHz, but for hi-fi music further refinement is required. Figure 8.25 shows a Musicam encoder which, by the use of several B channels, offers 20 kHz bandwith stereo transmission.

Computer applications

Using the plug-in cards, or terminal adaptors, a wide range of applications may benefit for the rapid call set-up and high throughput of the ISDN. For

Figure 8.24
Reporting via ISDN (photo courtesy Keith Hailey).

file transfer and electronic mail a change from a modem running at 9.6 kbit/s at 64 kbit/s is a sevenfold speed improvement. Of much greater interest are the new applications which were only rendered possible by the ISDN. One such is computer conferencing, shown in Figure 8.26. Here two or more people connected by an ISDN B channel can jointly run an application, amending the data (numerical or drawn) and when each is content with the outcome they can 'sign off' the result. The second B channel provides an audio link.

Another application (Figure 8.27) allows access to high definition pictures. The high data rate available, together with the sort of compression

Figure 8.25
Musicam stereo encoder (photo courtesy Corporate Computer Systems, Inc.).

techniques described in Section 8.4, allow the rapid downloading of such pictures as medical, artwork, maps and educational. The rapid call set-up means that a range of databases all over the world may be rapidly accessed; the PC in Figure 8.27 displays a menu of pictures available and selection is by mouse or keyboard.

A similar application is to use the network to access real time pictures at a multitude of sites for surveillance purposes. In Figure 8.28 the operator has a menu of telephone numbers in front of her, each of which has a slow scan surveillance camera connected. The attendant can view each camera

Figure 8.26
Computer conferencing (photo courtesy ICL).

Figure 8.27
Picture database access.

at will on the associated television screen. The system shown may access up to four cameras and display each on a quarter of the screen. This has obvious applications for security and traffic control.

Videophone

The ability to see as well as hear one's correspondent has been an engineer's ambition for a long time but attempts to do this in the 1960s in the USA by the setting up of special networks failed in the face of high cost and poor visual quality. Indeed there is a line of argument that the service is in fact undesirable on the grounds that a visual service will require you to get dressed (or at least tidier) before answering the phone! Nevertheless the ISDN does make a service feasible, using the techniques described in

Figure 8.28
Telesurveillance.

Figure 8.29
ISDN videophone.

Section 8.5. Figure 8.29 shows a stand alone videophone. Service can be provided over a single B channel using 48 kbit/s for video and 16 kbit/s for speech, or over two B channels either using 64 + 64 or 112 + 16 kbit/s for video + speech. As an alternative the videophone function may be incorporated into a personal computer. In this case it is customary to have the videophone picture as a window which may be moved and zoomed like any other window. The intelligence of the PC may also be used to control a multipoint video-conferencing environment (Figure 8.30) integrated with other facilities allowing the exchange of numerical and graphical information between the participants.

Facsimile

Group 4 facsimile machines are available and their characteristics are described in Section 8.3. Most Group 4 machines also incorporate the facility to revert automatically to Group 3 when they discover that they are connected to another Group 3 machine at the distant end. In the Group 4

Figure 8.30
Multipoint integrated conference

Figure 8.31
Group 4 facsimile.

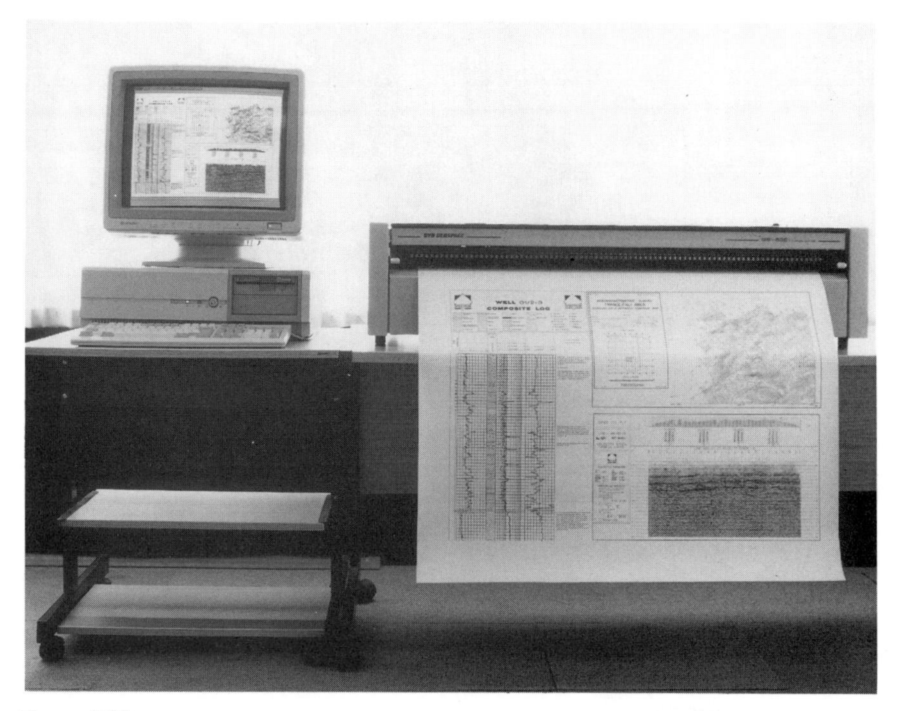

Figure 8.32
AO fax (photo courtesy Spectrum Energy and Information Technology Ltd).

mode the sending of documents of tens of pages of length is no hardship and so for this reason Group 4 machines may be differentiated in appearance mainly by the comprehensive paper feeding arrangements; see Figure 8.31.

With the higher transmission speeds colour fax is also possible, as are fax machines dealing with larger sheet sizes. Figure 8.32 shows a fax machine which handles the European A0 size sheet (about 47 × 33 inches, 1190 × 840 mm). At 64 kbit/s it may take 1 minute to transmit a sheet, compared with 25 minutes at its prevous 2.4 kbit/s speed.

Switches

Many PABXs are available which can interface to the ISDN. The larger PABXs are more appropriately connected to primary rate accesses, and the smaller key systems will use a few basic rate accesses. Early PABXs tend to use analogue or proprietary digital access to the extensions, but with the advent of low cost 1.420 devices the trend will be towards the use of standard ISDN interfaces for extensions. These switches may well be used for data purposes as well as speech and offer an alternative to local area networks. As described in Section 6.4 the term 'PTNX' may be more appropriate for such equipment.

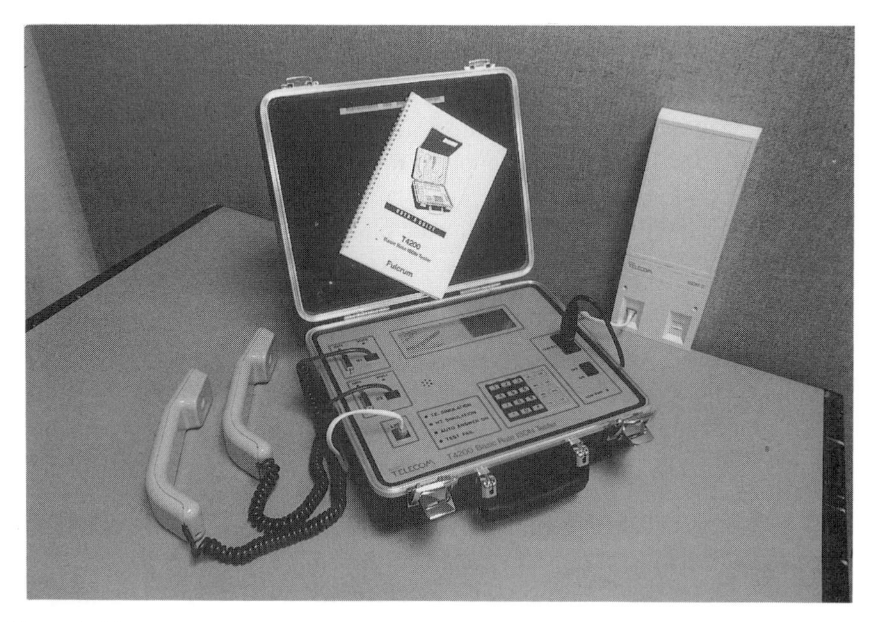

Figure 8.33
ISDN basic rate tester (photo courtesy Fulcrum Communications Ltd).

Testers

In surveying terminal applications and apparatus it may seem odd to include testers. However, they connect at the same point to the network and offer a service to the user in exactly the same manner that any other terminal apparatus does. The difference is that the user is normally a member of the network provider's organization. Testers must firstly determine whether the network is working, and secondly locate the problem if it proves that a fault exists. On the other hand, the tester must be inexpensive enough to be made widely available, and it must be simple to use. The tester shown in Figure 8.33 can set up and receive both data and telephony calls on both B channels simultaneously. Voice service may be demonstrated via the handsets and error monitoring may take place on data calls. Signalling messages are displayed. These message are also recorded so that if a fault is identified the messages may be 'replayed'; this is an important feature as the interchange of messages is normally much faster than the eye can observe and brain can interpret.

REFERENCES

CCITT Recommendations:

I.241	Teleservices supported by an ISDN.
G.721	32 kbit/s adaptive differential pulse code modulation (ADPCM).
G.722	7 kHz audio coding within 64 kbit/s.
G.725	System aspects for the use of the 7 kHz audio codec within 64 kbit/s.
T.4	Standardization of Group 3 facsimile apparatus for document transfer.
T.6	Facsimile coding schemes and coding control functions for Group 4 facsimile apparatus.
T.81	Encoding of Still Photographic Images.
T.411–418	Open Document Architecture (ODA).
T.431–433	Document Manipulation and Transfer protocol (DTAM).
T.503	A document application profile BTO for the interchange of Group 4 facsimile documents.
T.521	Communication Application profile BTO for Document Bulk transfer based on Session Service.
T.563	Terminal Characteristics for Group 4 facsimile apparatus.
X.224	Transport Protocol specifications for open systems interconnection for CCITT Applications.

X.225	Session Protocol specification for open systems interconnection for CCITT Applications.
H.221	Frame structures for a 64 to 1920 kbit/s Channel in Audiovisual Services.
H.231	64–2048 kbit/s multipoint control.
H.242	Communications procedures, 64–2048 kbit/s.
H.243	Communications procedures, 64–2048 kbit/s multipoint.
H.261	Video Codec for Audiovisual Services at p × 64 kbit/s.

International Standards Organisation

ISO 9282–2 Still picture compression technique

ISO 11172–3 Coding of moving pictures and associated audio for digital storage media at up to about 1.5 Mbit/s—Part 3 Audio

International Radio Consultative Committee

CCIR 601–1 Encoding parameters of Digital Television for studios

CEPT Recommendation

T/TE 06 series Videotex services presentation layer data services

QUESTIONS

1 Why is it necessary to define teleservices?

2 In an ADPCM system the decoding depends on the previous information received. How then are errors in transmission tolerated so that permanent errors of decoding do not result?

3 The high band signal in sub-band ADPCM encodes signals up to 7 kHz but is only sampled at 8 kHz. Produce a diagram like the one in the top right-hand corner of Figure 2.1 to show that the resulting sidebands do not overlap.

4 If a Group 4 fax machine is transmitting a page consisting of vertical, constant-width black and white stripes, what digital sequence will be generated to transmit this after the irst line (ignore end of lines)? What code sequences will be generated if the stripes are at 45° to the left and right?

5 Text displays on a screen may be encoded by the DCT or sent as International Alphabet 5 (ASCII) characters. By comparing the bits per pel required for each method, decide on the relative eficiency of the two for the purpose. Note that both forms of text appear in Figure 8.10 - look for the name on the back of the boat.

Chapter 9

Broadband ISDN

Up to this point we have only considered ISDN based on 64 kbit/s B channels and 16 or 64 kbit/s D channels. However, for many applications even higher bit rates would be useful. Although video telephones may be acceptable at 64 kbit/s using a very small screen, for video conferencing, where a normal television size screen is needed, 384 kbit/s (6 × 64 kbit/s) is more attractive. Entertainment TV must be able to cope with situations in which successive frames are very different to meet the artistic aspirations of producers. For this reason there is little in the way of redundancy reduction possible and bit-rates in the tens of megabit/s are needed. High definition television requires bit-rates in the hundreds of megabit/s.

On the business side the interconnection of high speed LANs used for computer-aided design may well generate traffic at high rates. Of course there is also the general drift of people's expectation. The ability to handle facsimile pages in 4 seconds compared with the 30 seconds of the pre-ISDN era may be widely appreciated, but how long will it be before people expect the fax machine to operate at the same speed as the office photocopier, and in full colour? Thus not only can the need be foreseen for higher rates, but also the technology to offer service is available in the form of optical fibre and devices.

Figures 9.1 shows how the data capacity of a public switched telephone network connection has increased as modems have developed and with the availability of the ISDN. The dotted lines give some plausible extrapolations for the future. The curved line indicates how modems have evolved under the constraint of Shannon's limit described in Section 1.4. The straight line offers a demand for the future without the Shannon limit. Figure 9.2 summarizes the data rate needs of various services. The provision of channels above 64 kbit/s is generally referred to as Broadband ISDN or B-ISDN but perceptions as to what constitutes a B-ISDN and the applications that must be carried vary enormously.

9.1 $N \times 64$ kbit/s

At the lowest end of B-ISDN comes the concatenation of several 64 kbit/s channels. CCITT Standard H.221 provides for the control and allocation of

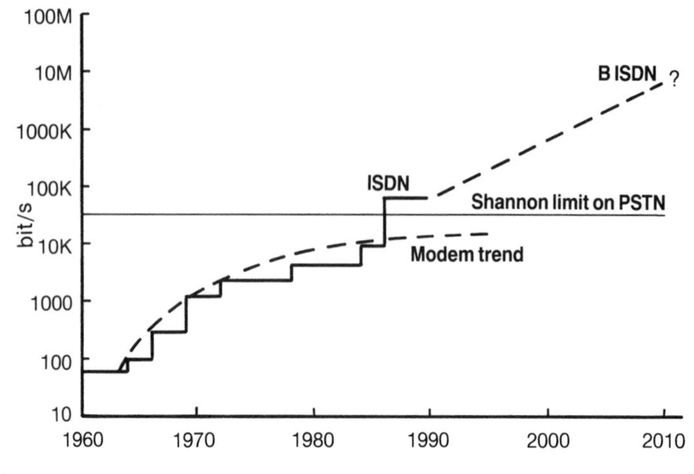

Figure 9.1
The rise of switched data rates.

bandwidth for services carried on such assemblies of 64 kbit/s channels. It is in this area that the BAS (see Section 8.6) relating to transfer rates and terminal capabilities comes into its own. For example a BAS of 001 01010 indicates a transfer rate of 384 kbit/s with 64 kbit/s allocated to audio information and 320 k/bits allocated to video.

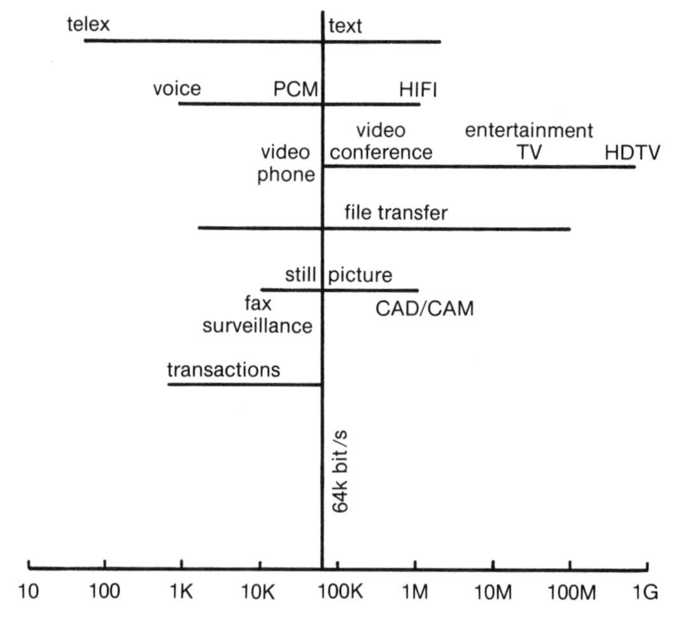

Figure 9.2.
Bandwidth requirements.

On the face of it the assembly of several 64 bit/s channels could be achieved on the ordinary ISDN by setting up several 64 kbit/s calls to the same destination on a primary rate interface and concatenating the channels at the terminal. Unfortunately the network does not provide for channels set up in such a way to be subject to uniform delay. Time switching stages (see Section 3.3) may introduce up to a frame's delay in one channel relative to another. More seriously there is no guarantee that channels will follow the same, or even similar, paths through the network. Different channels may be routed via different transmission line plant on different routes with different delays; the most extreme situation would be where one channel is carried by a satellite link and another is carried on a terrestrial link. The provision of alternative routing strategies in the network may also mean that one channel may pass through an extra switching node with its associated delays. There are two ways of overcoming these problems.

The terminal solution

Obviously by providing appropriate buffers at the terminals the delays in each channel can be padded to be equal. This will necessitate an initial 'investigation' period by the terminals to establish the delays. One assumption that may reasonably be made is that relative delays will not change during a call as, in general, networks do not normally reconfigure established calls. CCITT Standard H.221 includes strategies for establishing such connections. Bit 1 of frames 0, 2, 4, 6 in the multiframe are allocated for this purpose. The strategy is that these bits in each channel would contain a 4-bit binary number which is decremented on each multiframe. These would be synchronized in each channel at the sending end and the receiving terminal could then determine the difference in delay between the various channels. Other standard procedures are also being developed.

The network solution

In this case the exchange processors would ensure that all channels are kept within a single time division multiplex and hence follow a common route. Apart from the need for additional software, the major problem is in design for traffic carrying capacity. Routes are dimensioned to give an acceptable quality of service when channels are required singly. For example using Erlang's formula (implying the usual assumptions) a group of 30 or 23 trunks could be expected to carry 19.03 or 13.42 respectively erlangs of traffic and only 1 in 200 calls would be unable to find a free channel (i.e. a grade of service of 0.005). For a channel operating at this traffic level

Table 9.1 shows the probability of not being able to accept a multichannel call of N channels. This is the Grade of Service for $N \times 64$ kbit/s circuits in an environment where multichannel calls are rare (and hence the chance of two arising is negligible) and single 64 kbit/s channels predominate in the design consideration. Reading from the lower rows of these tables, the probability of not being able to find 6 channels free on a 30 channel system is 0.102 – i.e. about 1/20 of the basic Grade of Service. With 23 channel systems the same probability rises to 0.129 – i.e. about 1/26 of the basic grade of service. The Grade of Service offered for single 64 kbit/s channels, after setting up the 6×64 kbit/s channel, is to be found in the middle rows (i.e. 0.050 or 0.072 for 30 and 23 channels respectively). For these reasons it might be felt that $N = 6$ is about the largest multichannel service that could be offered in this way although even this might need a special provision. Of course in an exchange and trunk network there will be more than one 30- or 23-channel group from which to choose to give the N channels but on a single primary rate ccess there is no advantage from this source.

Table 9.1
P_{+1} is the grade of service for single channels while an N-channel call is in progress
$P_{<N}$ is the probability that fewer than N channels will be free.

30-channel system

	0	1	2	3	4	5	6	7	8	9	10	11	12
N =	0	1	2	3	4	5	6	7	8	9	10	11	12
P_{+1} =	0.005	0.008	0.012	0.018	0.026	0.037	0.050	0.066	0.086	0.108	0.134	0.163	0.195
$P_{<1}$ =		0.005	0.013	0.025	0.042	0.067	0.102	0.147	0.203	0.272	0.351	0.438	0.530

23-channel system

	0	1	2	3	4	5	6	7	8	9	10	11	12
N =	0	1	2	3	4	5	6	7	8	9	10	11	12
P_{+1} =	0.005	0.009	0.014	0.023	0.035	0.050	0.072	0.098	0.129	0.166	0.207	0.253	0.303
$P_{<N}$ =		0.005	0.014	0.028	0.050	0.083	0.129	0.191	0.270	0.365	0.470	0.580	0.686

9.2 H CHANNELS

Referring to Figures 9.1 and 9.2, one might identify some specific steps in the rates offered above 64 kbit/s. These have been identified by CCITT as:

(a) H0 at 384 kbit/s (i.e. 6×64 kbit/s). This would be specifically attractive for video conference codecs and hifi sound. The potential effect of this on traffic design has already been discussed in Section 9.1. It is assumed that the interface to the customer for H0 channels will be the primary rate interface. The 1.544 Mbit/s primary rate can accommodate three H0 channels in timeslots 1–6, 7–12, 13–18. If a signalling channel is not required then a fourth H0 channel can be accommodated in timeslots 19–24. At 2.048 Mbit/s five H0 channels can be accommodated and the preferred assignment is to use channels 1–6, 7–12, 13–19 (excluding 16), 20–25, 26–31. Any timeslots not used for H0 channels can be used for B channels.

(b) H1 channels. Two forms of H1 channel have been identified:

H11 at 1.536 Mbit/s. This may be carried on a 1.544 Mbit/s primary rate interface but a signalling channel would have to be provided separately. Alternatively it can be carried in timeslots 1–25 (excluding 16) of a 2.048 Mbit/s primary rate interface. Signalling can then be carried in channel 16.

H12 at 1.920 Mbit/s. This may be carried in timeslots 1–31 (excluding 16) of a 2.048 Mbit/s primary rate interface. Channel 16 is used for signalling as usual.

Because of the large chunks of capacity that H1 channels would absorb in the switching fabric of a switch based on the switching of 64 kbit/s channels, it is most likely that either a new switching network will be provided or special provision will be made within switches for them.

(c) CCITT has also identified some other H channels: H21 around 34 Mbit/s, H22 around 55 Mbit/s, H4 around 135 Mbit/s.

9.3 HIGHER RATE INTERFACES

Optical fibres offer virtually unlimited bandwidths and their use in the local network is widely predicted. Given the demand the technical and economic problems of providing service to customers can undoubtedly be overcome. Even if it is unacceptable to use one fibre per connection, then wavelength division multiplexing offers enormous capacity, particularly if associated with coherent detection techniques.

Various channel rates have been identified above, but it is not necessary that there are interfaces to the customer operating at that rate. The whole concept of the ISDN is that there should be a minimum number of interfaces identified so that a wide range of compatible equipment should become available. So far only two interfaces have been identified—the basic rate and the primary rate (although this comes in the North American and Rest of World versions). Continuing to higher rates, once again there is pressure to have only two more interfaces; the proposal is that these should operate at about 150 and 600 Mbit/s. This does not mean that only channels at these rates would be available. In fact, as will be seen, the plan is that many channels of a wide range of rates will be multiplexible on to these interfaces, so that a wide range of terminal equipment can be connected to them. At the 150 Mbit/s rate it can be seen that all services except high definition TV can be accommodated. 600 Mbit/s could accommodate several standard TV channels or HDTV, simultaneously with lower rate services. When studying Figure 9.2 it is important to appreciate that the horizontal scale is logarithmic; on a linear scale the small capacity required for the lower rate services would be more easily appreciated because the ranges would be invisible on the left-hand side. The real problem is how services with such a wide range of rates could be efficiently multiplexed on to a common bearer and two solutions are being followed. One is based on synchronous multiplexing, but with a format which can be configured to match the needs of the user under his control; this is called the Synchronous Digital Hierarchy (SDH). The other extends the use of packed mode services to a very lightweight protocol called Asynchronous Transfer Mode (ATM) which because of its simplicity can be implemented at these high speeds.

9.4 SYNCHRONOUS DIGITAL HIERARCHY (SDH)

Paul McDonald

A new international multiplexing standard, the Synchronous Digital Hierarchy (SDH), is currently being adopted by many networks for their high speed transmission networks. SDH evolved from the American optical interface standard, SONET, which was designed to solve the shortcomings of the transmission hierarchy currently in use, the Plesiochronous Digital Hierarchy (PDH). After modifications to accommodate European interface rates, it was adopted by the CCITT as a worldwide transmission Standard, and is detailed in the CCITT Recommendations G.707, G.708, and G.709.

The current PDH transmission network multiplexes channels into higher bit-rate structures on a stage-by-stage basis, with each stage using its own multiplexing and framing methods. As a result it is very difficult to access individual channels within a high bit-rate signal. Converting between different channel rates, removing or adding individual channels generally requires the high rate signal to be totally demultiplexed to gain access to the constituent channels. This requires a complete range of multiplex equipment to be used, resulting in complexity and cost. Figure 9.3 shows an example of the various multiplexing stages required by the North American and European PDH (note: the 140 Mbit/s interface is rarely used in North America).

In contrast SDH provides a considerably improved method of multiplexing channels into high bit-rate interfaces of 150 Mbit/s and above. Com-

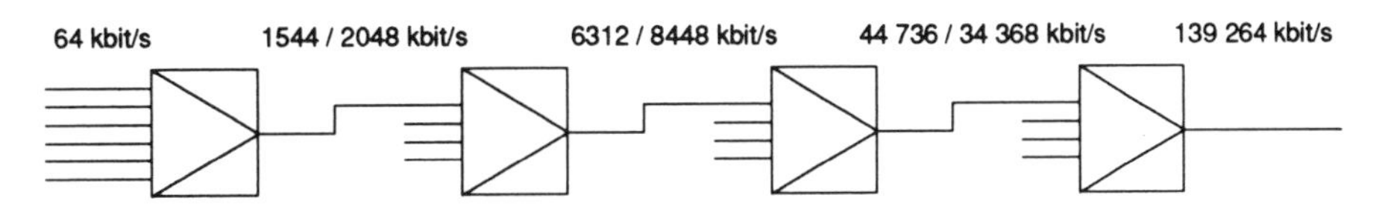

Figure 9.3
PDH multiplex structure in North America/Europe.

pared to the PDH, SDH is able to offer:

— One worldwide standard for multiplexing and interworking.
— Direct access to lower rate channels without having to demultiplex the entire signal.
— Simplified evolution to higher bit-rates.
— Comprehensive provision for network management.
— Interconnection between independent networks without introducing frame slips.
— The ability to carry new broadband channels as they appear, such as the transport of ATM based services.

9.4.1 Virtual containers

In the SDH standards the transmission network is segregated into regenerators, multiplexes, and the functions required to transport information between the end points of the network. The regenerator and multiplexing functions are handled by the section layers of the protocol, the end-to-end functions are provided by the protocol's path layer.

This path layer transports the information across the network by encapsulating it inside structures called Virtual Containers (VC). These virtual containers consists of two parts:

(1) The container (C) which holds the data to be transported.
(2) The Path Overhead (POH), which provides maintenance channels and control information that is associated with the path across the network.

Different classes of virtual container have been defined, each designed to transport the various channel types currently found in the network. Table 9.2 shows the different types of virtual container that are available.

Information inside a virtual container is normally carried transparently by the SDH network. However, the VC-1s can also carry channels that are synchronized to the SDH data stream. This gives the SDH multiplexes direct access to the individual 64 kbit/s channels within the VC-1.

Table 9.2
Virtual container sizes.

Virtual Container	Container Capacity	Services Supported
VC-11	1.7 Mbit/s	1.544 Mbit/s North American channel rates
VC-12	2.3 Mbit/s	2.048 Mbit/s European channel rates
VC-2	6.8 Mbit/s	6.312 Mbit/s channels (rarely used). VC-2s can also be concatenated together to carry higher rate services
VC-3	50 Mbit/s	34.368 Mbit/s and 44.736 Mbit/s channels
VC-4	150 Mbit/s	139.264 Mbit/s channels and other high bit-rate services

9.4.2 SDH frame structure

The SDH signal is constructed from STM-1 frames (Synchronous Transport Module level 1). This signal provides an interface bit-rate of 155.52 Mbit/s and it is the basic unit from which higher interface rates are constructed. For these higher rates N STM-1 frames are byte interleaved to form an STM-N signal. The value of N=1, 4, and 16 have been standardized by the CCITT, giving interfaces of 155.52 Mbit/s, 622.08 Mbit/s, and 2.488 32 Gbit/s respectively.

Each STM-1 frame consists of 2430 bytes transmitted every 125 μs. Due to the large number of bytes the frame is normally depicted as a rectangular structure comprising of 9 rows by 270 columns of bytes, as shown in Figure 9.4. This structure is transmitted form top left to bottom right, scanned in a similar method as reading a book.

The first 9 columns of the frame, except row 4, are used for section overhead functions. The regenerator overhead is used to transfer the signal

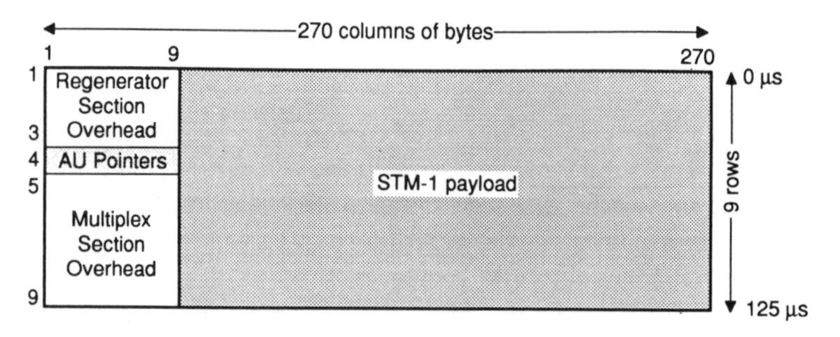

Figure 9.4
STM-1 frame structure.

between line equipment and it contains facilities for framing, error detection, and management communications channels. The multiplex section overhead is used between the multiplexes, providing facilities for block error detection, automatic protection switching, and a 576 kbit/s data communications channel.

The rest of the available capacity is used to carry the payload. This is done using Administrative Units (AUs) which provide a mechanism for the efficient transportation of payloads that are not in exact synchronization with the STM-1 signal whilst avoiding frame slips.

The administrative unit operates by transporting information inside a virtual container. This is allowed to 'float' inside the payload area of the STM-1 frame. The start of this virtual container is indicated by an administrative unit pointer which contains the distance in bytes to the start of the next virtual container. Thus the STM-1 frame and the VC can have different phases, and slight differences in the frequency of the two signals can be accommodated by moving the starting position of the VC within the frame. An example of the AU-4 structure, which utilizes all of the capacity of the STM-1 payload area, is shown in Figure 9.5.

The AU pointer consists of two parts; the offset in bytes to the next virtual container, and a negative justification opportunity which normally holds dummy data. If, however, the VC data rate is slower than the STM-1 frame rate, then this justification opportunity is occasionally filled with data from the VC. This will move the end of the VC, and thus the start of the following VC, earlier in the STM-1 frame. This change is reflected in the next AU pointer by decrementing its value. Similarly if the VC is slower than the STM-1 frame, then the VC is delayed by inserting dummy data

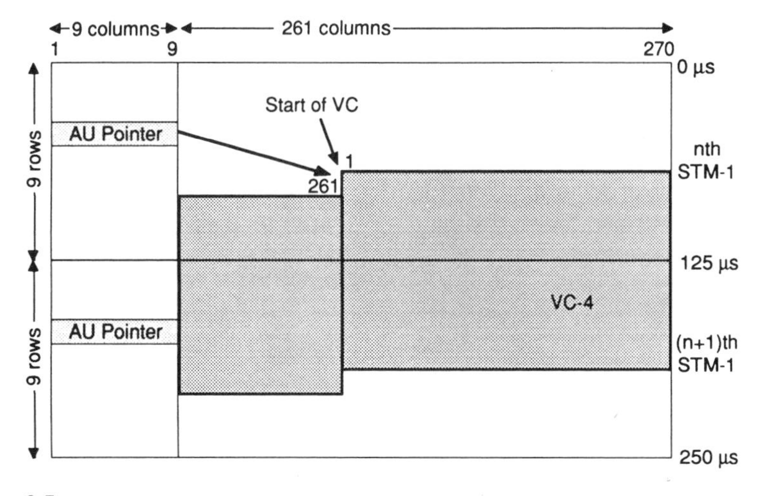

Figure 9.5
VC-4 floating inside STM-1 frames.

into the bytes following the AU pointer. Thus the end of that VC will occur later in the STM-1 frame, delaying the following VC, and the value of the next AU pointer is incremented. In both cases the size of the VC does not alter, just the time taken to transmit it.

Two different types of administrative unit are defined—the AU-4 and the AU-3.

AU-4

The AU-4 transports a single VC-4 which occupies the whole of the STM-1 payload area. As the VC-4 can 'float' inside the STM payload area it is normally illustrated as shown in Figure 9.6. This VC-4 has the capability to transport 139.264 kbit/s channels and it has 9 bytes of path overhead available. The single AU pointer required by the VC is held in row 4 of the STM-1 frame.

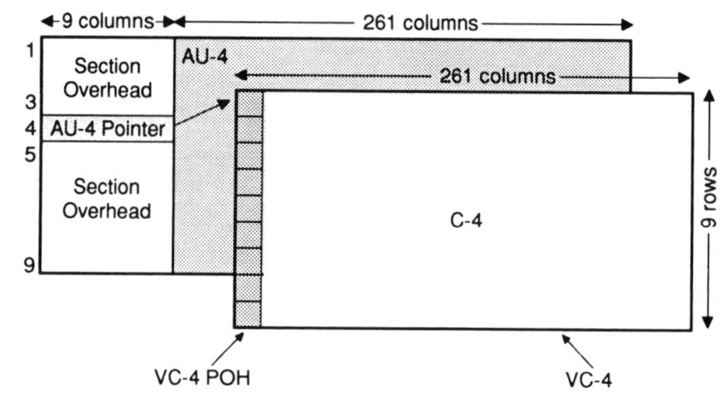

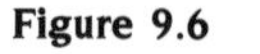

Figure 9.6
VC-4 floating in an AU-4.

Higher rates signals, in an STM-*N* interface, can be carried by concatenating several VC-4s together to form a larger VC. The individual VC-4s of this concatenated signal are treated as sone unit for frequency justification operations.

VC-3

For lower rate signals, around 50 Mbit/s, the AU-3 structure is defined. This allows three independent VC-3s to be transported within the STM-1 payload area, each with its own AU pointer. This structure is illustrated in Figure 9.7.

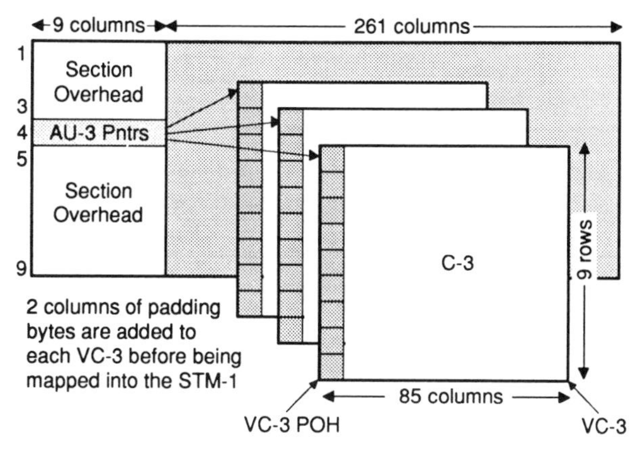

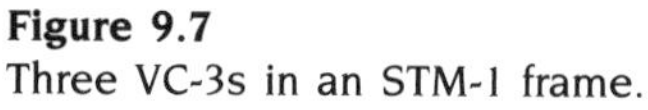

Figure 9.7
Three VC-3s in an STM-1 frame.

When the three VC-3s are multiplexed into the STM-1 payload area they are byte interleaved so that a byte of the first VC-3 is followed by a byte of the second VC-3, then the third. Each of the VC-3s has its own AU pointer. This allows the start of the VCs to be different from each other so that they can 'float' independently within the STM-1 frame. Like the VC-4, the VC-3s each have 9 bytes of path overhead available. However, when each VC-3 is multiplexed into the STM-1 frame two columns of padding bytes are added to fill out the available capacity. These padding columns are added between columns 29 and 30, and columns 57 and 58 of the VC-3.

9.4.3 Tributary units

VC-3s and VC-4s can be transported by the STM-1 frame using the AU mechanism detailed above. Lower rate VCs require an additional step to permit the mixing of different types of lower rate VC in the frame, and also to accommodate VCs originating from different sources.

Lower rate VCs are first placed inside structures called Tributary Units (TUs) before being placed inside a larger VC-3 or VC-4. The tributary unit is similar to the AU mechanism. TU pointers permit lower-order VCs to 'float' independently of each other and of the higher order VC transporting the TUs. This is shown schematically in Figure 9.8.

A VC-4 can carry three VC-3s directly, using a TU-3 structure similar to the AU-3. However, the transport of VC-1s and VC-2s inside a VC-3 or VC-4 is more complicated. An additional step is required to simplify the process of mixing the different types of VC-1 and VC-2 into the higher order VC.

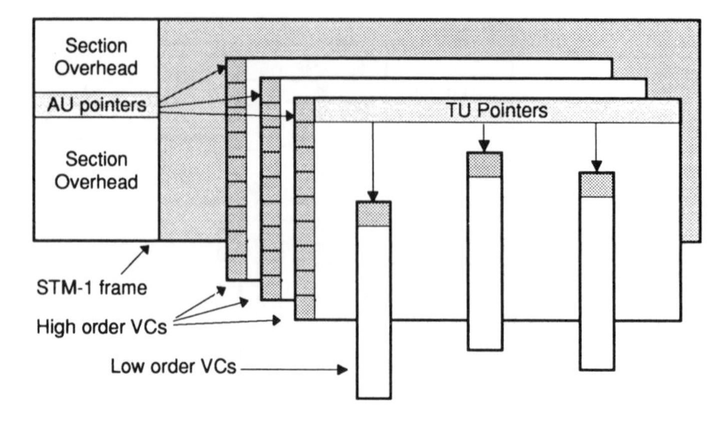

Figure 9.8
Transport of lower rate VC-s using the TU structure.

Once the VC-1s and VC-2s have been formed into their TUs, these are placed inside a Tributary Unit Group (TUG). This TUG-2 is a 9 row by 12 column structure which can contain either four VC-11s, three VC-12s, or a single VC-2. Each TUG-2 can only contain a single type of VC, but this TUG-2 can be freely intermixed with TUG-2s containing other VC types. The fixed size of the TUG-2 removes the differences between the sizes of the VC-1s and VC-2s, thus making it easier to multiplex mixtures of VC types inside the higher order VC. Throughout the process of multiplexing TUs into a TUG-2, and TUG-2s into a VC-3 or VC-4, byte interleaving is used to minimize buffering delays. An example of VC-1s and VC-2s being placed inside a higher order VC-3 is shown in Figure 9.9.

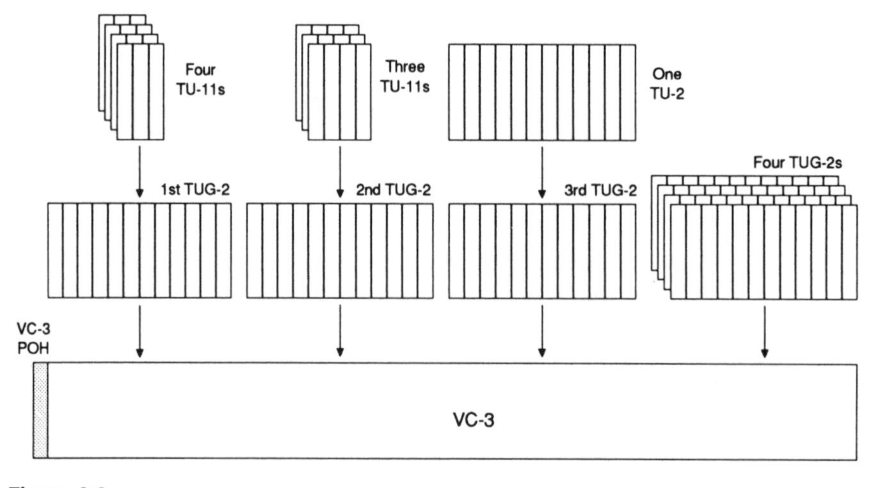

Figure 9.9
Placing TUGs inside a VC-3.

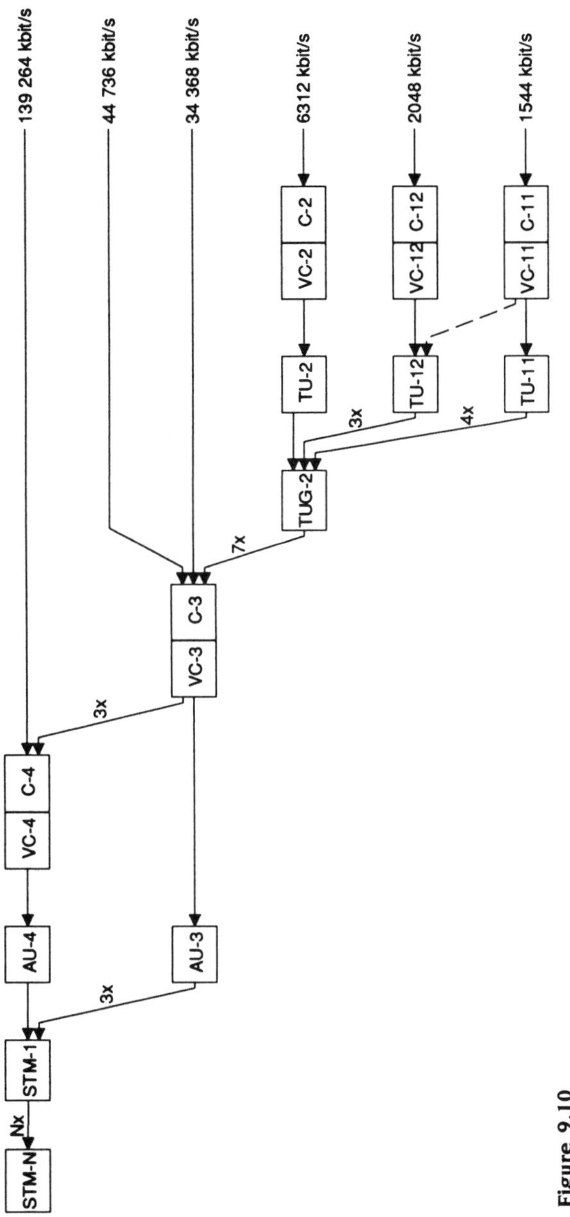

Figure 9.10
SDH multiplex structure.

The above description has shown that several steps are required to multiplex low bit-rate channels into the high bit-rate SDH interface. These steps are summarized in Figure 9.10. In North America lower order VCs are transported using VC-3s. In Europe the lower order VCs are carried directly by a VC-4, and the VC-11 is carried using the TU-12 mechanism (with the addition of padding bytes).

Although the multiplexing mechanisms used appear to be complicated, in reality it is not too difficult for the multiplexing equipment to locate the position of the individual low bit-rate streams using the pointers. This results in a large decrease in the cost and complexity of the multiplexing equipment compared to the PDH. Also the ability of VCs to 'float' independently simplifies the process of removing one channel and replacing it with another channel originating from a different source, permitting 'add-drop' multiplexes to be implemented economically. Additionally SDH provides a simple method of extending the interface to bit-rates of 622 Mbit/s or higher.

9.5 ASYNCHRONOUS TRANSFER MODE (ATM)

Richard Boulter

The previous section outlined a method of synchronously but flexibly multiplexing together channels of a range of rates. However, it may be deduced that even in outline the overall process turns out to be complex. A full description of the processes involved is several times the length allocated to Section 9.4. However, the complexity primarily lies in the control mechanisms which are required to operate at low speeds of a few kbit/s rather than the interface rate.

An alternative strategy is to adopt message based techniques of the type described in Chapter 7. Using such techniques it is straightforward to accommodate a wide range of channel rates on a single bearer. The penalty is that much of the control mechanism has to operate at the bearer rate. However, present day technology only allows the simplest processing activities at rates of 150 Mbit/s, and at 600 Mbit/s even the simplest of operations are challenging. The use of full packet switching techniques would be prohibitive.

ATM uses short, fixed-length cells with minimal headers, to allow calls to be routed at high speed by means of hardware-implemented routing tables at each switch. International agreement is that the header of each cell will consist of 5 bytes and that the cell information field will consist of 48 bytes, making a total cell length of 53 bytes. Note the basic unit is called a 'cell' to avoid confusion with 'packets' of lower bit-rates. The header will consists of bits to provide two main routing functions.

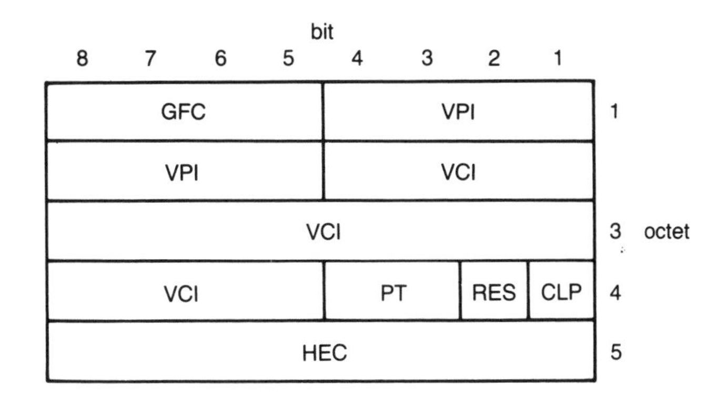

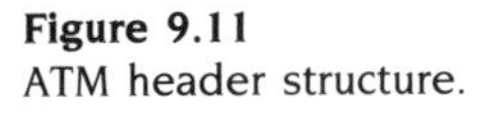
Figure 9.11
ATM header structure.

(a) *Virtual Path Identifier (VPI).* A path is the equivalent of a route in a circuit based environment, permanently connecting two points together. In an ATM environment the path would not have a fixed capacity. The 'virtual' tag indicates that cells would be routed from node to node on the basis of the VPI, the route being established at the beginning of each call on the basis of signalling messages.

(b) *Virtual Call Identifier (VCI).* These calls would be set up as required over the virtual path indicated by the VPI.

The structure of the header at the user–network interface is shown in Figure 9.11. It contains the following sub-fields:

— Generic flow control field (4 bits).
— Routing field (VPI, VCI); 24 bits are available.
— Payload Type (PT) field (2 bits).
— Reserved field (1 bit)
— Cell Loss Priority (CLP) indicates whether the cell has a lower priority and can therefore be discarded during overload conditions.
— Header Error Control (HEC) field (8 bits).

The ATM network would be able to flexibly handle all types of traffic. Particular examples are:

(a) *Voice.* 64 kbit/s voice can be assembled into ATM cells. With 48 bytes of information in each cell then each cell can contain 6 ms of speech. For some purposes the delay this introduces may be excessive and only partially filled cells may be used.

(b) *Music.* The Musicam coding process described in Section 8.2 inherently generates a variable bit-rate and so is appropriate for ATM transmission.

(c) *Video*. As discussed in Section 8.5, the main technique for reducing the capacity required for a video signal is to transmit only information relating to the changes in picture content. This varies with the action in the picture and so is also ideally suited to the variable capacity of ATM.

(d) *Signalling*. This would be based on the standards for narrow band ISDN, but assembled into cells. Some of these cells may have to have high priority (e.g. flow control) particularly under congestion conditions.

The services do not have to produce information at a constant rate either, which will be more efficient when transmitting interactive data or using video coding techniques where only changes in picture content have to be transmitted. It is for these reasons that ATM is seen as an ideal candidate for an integrated broadband network.

9.5.1 Reference models

The reference model established for narrow band ISDN is considered sufficiently general to be applicable to the broadband environment. The reference points and functional groups are therefore appended with the letter B as shown in Figure 9.12 to indicate their use for broadband. The broadband network termination (B-NT1) terminates the transmissions system used to provide access to the first switching node and supports the interface to either another broadband network termination (B-NT2) or, if this is not present, to a broadband terminal directly. The B-NT1 also

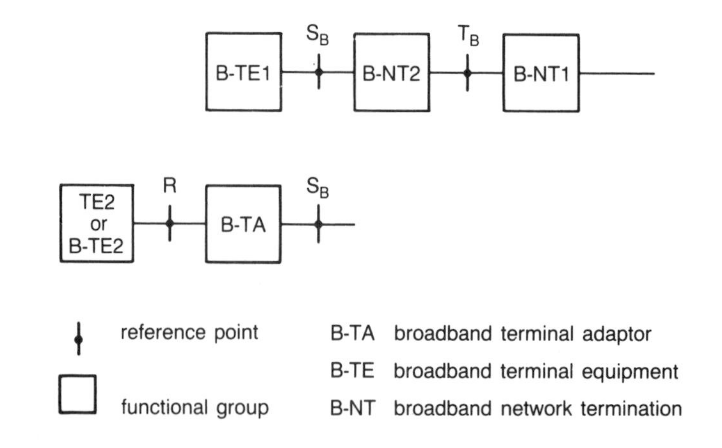

Figure 9.12
B-ISDN reference configurations.

provides maintenance features for use by the public network operator and for this reason in Europe is seen as part of the public network. The more complex B-NT2 often contains a local switching element for a customer premises network.

The interface at the T_B reference point is regarded as the regulatory interface in Europe and will provide a broadband interface to a broadband customer premises network (the B-NT2 function) which itself will then provide access to broadband terminals via the interface at S_B. The relationship of the interfaces at these two reference points is one of the key issues, including whether the interfaces should be point-to-point or point-to-multipoint. If the B-NT2 function is not always present, terminals will need to be connected directly to the B-NT1 function. This would seem likely in the case of small domestic customers who may only have the need for one broadband terminal. To ensure terminal portability the interfaces at T_B and S_B reference points must therefore be compatible. However, since interfaces at S_B are likely to be in the majority any additional features required at T_B must not impose any additional cost of the interface at S_B.

Since the interfaces have to support services varying from telemetry requiring a few bit/s up to high definition services requiring sevral hundred Mbit/s there is some support for a range of interfaces at S_B. However, to ensure terminal portability this should be discouraged. The major reason for having a number of interfaces is to prevent high volume (perhaps low speed services) having to bear any additional cost of a lower volume, high speed service. However, if the number of interfaces is restricted to one or possibly two, then the high volumes required will keep the cost of the interface down.

It is generally agreed that because of the high speeds involved, point-to-multipoint at the physical media dependent layer of the interface will be very difficult over the range of lengths of cabling required. However point-to-multipoint at the higher layers of the interface is possible. One way to provide the equivalent of a point-to-multipoint interface is to cascade the terminals in series as shown in Figure 9.13 making each terminal support an interface socket, equivalent to that at the B-NT2 or B-NT1. More work,

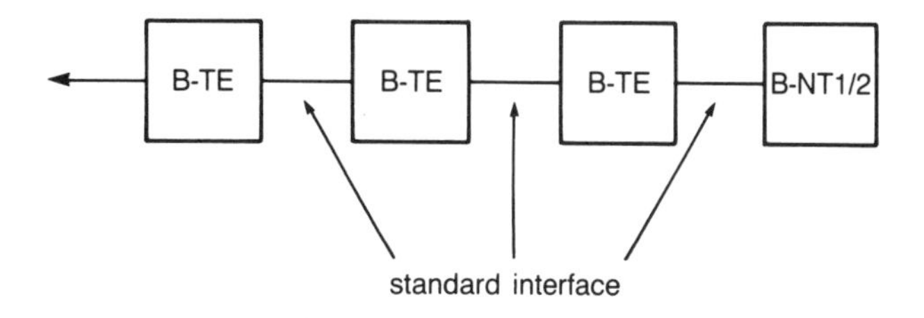

Figure 9.13
Cascaded terminals.

however, needs to be done on how the multipoint approach is tackled above the physical media.

9.5.2 Local access

The transmission link between the customer's premises and the local exchange, i.e. between the B-NT1 and the first switching node, is the U interface. In the narrow band ISDN environment this is often provided by the copper pair, but in the broadband environment, with its greater rates, coaxial cable, optical fibre or even satellite links will be required. As well as the possibility of using various physical media, different multiplexing techniques could be used. For example, in the short term the early plesiochronous multiplexing structure may be used but this is expected to be superseded by the synchronous digital hierarchy described in Section 9.4. The access could also be provided by employing passive optical devices in the network with time division multiplexing (TDMA) equipment or even a cellular radio network with restricted throughput. One of the main functions of the B-NT1 is to provide this interworking between a common user–network interface and this variety of network transmission systems and configurations. This will thus ensure terminal portability between different networks. In addition to the basic physical and transmission conversion, the B-NT1 may include rate conversion. The full transmission rate may be supported on the user–network interface, but the transmission into the network may be retricted by, for example, a radio link. In these circumstance there would be a need for flow control as well across the interface.

9.5.3 The user–network interface

In the UK and Europe this interface is likely to represent the regulatory interface between the liberalized Customer Premises Equipment (CPE) and the boundary of the Public network Operator (PTO) as is the case for narrow band ISDN. In narrow band ISDN two well-defined interfaces operating up to 64 kbit/s support a wide variety of CPE and applications which are able to evolve independently of the PTO network. A similar well-defined set of interfaces is therefore required for broadband ISDN and considerable effort is being put into this activity within the Standards bodies of CCITT and elsewhere. To help in this specification a number of protocol reference models have been developed.

9.5.4 The protocol reference models (PRM)

In CCITT an abstract 7-sub-layered architecture has been developed, covering roughly the OSI Layers 1 to 3, as shown in Figure 9.14. The main interest at present is in the network layers and the model shown in Figure 9.15 has evolved to show an ATM layer and an ATM adaptation layer above the physical layer but below Layers 2 and 3 (the higher layers) of the

	control plane	user plane
SL3	sublayer 3	sublayer 3
SL2	sublayer 2	sublayer 2
ADP	adaptation sublayer	adaptation sublayer
ATM	ATM sublayer	
SYN	synchronous channel/synchronized cells sublayer	
PFR	periodical physical frame sublayer	
PHY	physical medium dependent sublayer	

Figure 9.14
A generic layering architecture.

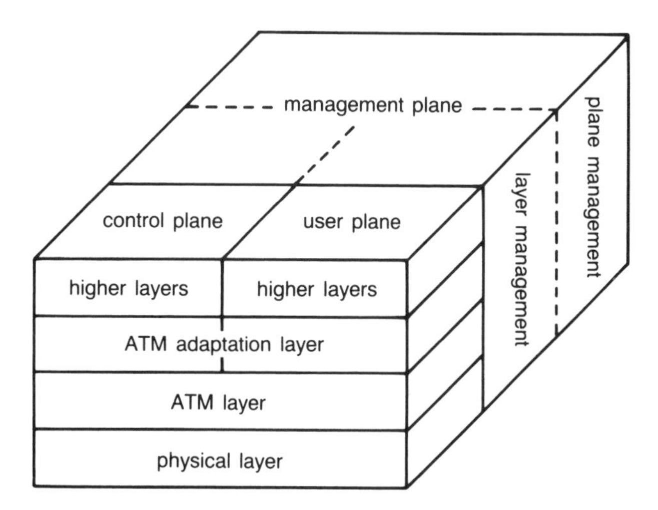

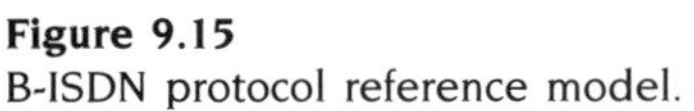

Figure 9.15
B-ISDN protocol reference model.

OSI model. Separate user, control and management planes have also been identified to reflect possible different procedures in these planes (described in CCITT Recommendation I.321). The main features of these layers are summarized in Figure 9.14.

Physical layer

Two interface rates have been defined and these have been chosen to align with the 155.52 Mbit/s rate of the Synchronous Digital Hierarchy (SDH) at Level 1 (STM-1) and the 622.08 Mbit/s rate at Level 4 (4 × STM-1). These high rates were chosen because of the perceived need to be able to support digitally encoded high definition TV signals. Not all of this capacity will be available to the user, since some is required to carry the ATM headers as well as any maintenance overheads that are required. In fact less than 135 Mbit/s is available to the user directly.

In the same way that a range of different media are available in the local access network, different media such as physical pairs, coaxial cable and optical cable, both multi- and monomode fibre, were investigated for the physical layer. However, only coaxial cable and optical fibres seemed to satisfy the performance criteria of both interfaces. In the short term it can be shown that two coaxial cables are more cost effective for the 155 Mbit/s interface if a reach of only 100 m is required. However, if greater reaches up to 2 km are required then optical fibre should be used and its cost is expected to fall in the longer term as new techniques are developed, eventually making it as cost effective as coaxial cables.

Two line codes were considered for the interface—Coded Mark Inversion (CMI) and a set–reset scrambler. The CMI coding scheme transmits a binary 0 as a half-width negative-going pulse followed by a half-width positive pulse. A binary 1 is coded either as a full-width negative pulse, or a full-width positive pulse, in such a way that the level alternates for successive binary 1s. Thus CMI coding provides a balanced line code which produces a line signal with no DC component and little low frequency content whilst retaining high clock frequency content which gives easy clock recovery. Its main disadvantage is that the baud rate is twice the information rate and therefore the transmission range is reduced. The scrambler technique on the other hand does not increase the baud rate but its effectiveness depends upon the statistics of the data to be scrambled. The CMI line code is preferred for the 155 Mbit/s interface although alternatives are being considered for 622 Mbit/s.

The physical layer also includes the framing structure. Two alternatives have currently been identified, the cell-based structure and the SDH-based structure. The first consists of a continuous stream of cells each containing 53 octets; in the second the stream of cells are mapped into the payload of

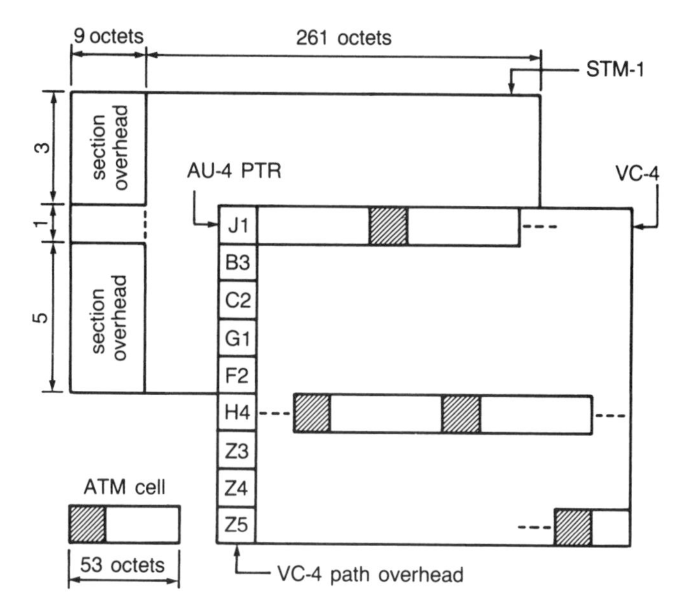

Figure 9.16
Mapping for ATM cells into the STM-1 frame.

an STM-1 frame of the SDH as shown in Figure 9.16. Maintenance messages are contained within the section and path overheads of the SDH frame in the latter case but special operation and maintenance cells have been defined for inclusion in the former such that they can easily be recognized at the physical layer from a unique pattern in the header.

Other issues to be considered at the physical layer include transmission performance (jitter, EMC, etc) and power feeding (activation/deactivation, emergency services and maintaining service in the event of a power failure).

ATM layer

This layer specifically addresses the ATM cell structure and coding. As shown in Figure 9.12 the cell is composed of a 5 octet header and a 48 octet information field. The primary role of the header is to identify cells belonging to the same virtual channel in the ATM stream.

ATM adaptation layer

This layer, as the name implies, adapts and maps the higher layers supporting a range of different services onto the common ATM layer. Since

	Class A	Class B	Class C	Class D
Timing relation between source and destination	Required		Not required	
Bit rate	Constant	Variable		
Connection mode	Connection-oriented			Connection-less

Figure 9.17
Service classification for adaptation layer.

difference services have different characteristics and are carried in different ways by these higher layers, then they will each require different ATM adaptation layer (AAL) protocols if they are all to be carried in an optimized manner. However it is hoped to restrict the number of AAL protocols with minimal effect on the services. Currently four classes of service have been defined based upon the timing relation between source and destination, whether there is a constant or variable bit rate and whether it is a connection oriented or connectionless oriented signal. These four services, which are summarized in Figure 9.17, will each then have a different AAL protocol.

In each of these AALs, two sub-layers have been defined, the segmentation and reassembly sub-layer (SAR) and the convergent sub-layer (CS). The SAR divides the higher layer information to be carried by the ATM layer into segments suitable for carrying in the 48 octet information field of the ATM cell and vice versa. The CS is service dependent and handles all other functions such as cell delay variations, cell loss, timing and error monitoring and control.

Higher layers

A common ATM bearer service supports the higher layers in both the user, control and management planes via the service dependent adaptation layer protocol. The higher layers of each service or application in the user plane will have different specifications enabling new services and applications to be defined at the higher layers yet still be supported by the common ATM bearer service. The higher layers of the control plane are initially to be based upon the narrow band ISDN signalling messages and protocols of CCITT Recommendation I450/1 (Q930/1) and thus will be different again to those of the user plane.

9.5.5 Signalling

A meta-signalling virtual circuit is always present over which a defined metal-signalling protocol is carried. This protocol is used to establish signalling virtual channels (SVC) as required over the user/network interface between the CPE and the network. Any number of SVCs can be established and they are individually identified by signalling virtual channel identifiers (SVCI) which are assigned and removed as required by the meta-signalling channel protocol. When the CPE has established a SVC then it is used to support the higher layer signalling messages and protocols.

9.6 THE FUTURE

The aim of telecommunication has long been to simplify its use to the same level as that required for electricity supply. For domestic and office environment power comes from a socket which is suitable for a wide range of equipment ranging from a clock consuming less than a watt, up to a kettle consuming a kilowatt or two. It has long been held as a dream that a similar arrangement could exist for telecommunications—a universal socket which could be used for anything from low speed data for telemetry and control, right up to high definition television. The promise of broadband ISDN is that this vision will become reality. The only question is when?

REFERENCES

CCITT Recommendations:

G.707	Synchronous digital hierarchy bit rates.
G.708	Network node interface for the synchronous digital hierarchy.
G.709	Synchronous multiplexing structure.
H.221	Frame structue for a 64 kbit/s channel in audiovisual teleservices.
I.113	B-ISDN Vocabulary.
I.121	Principles of B-ISDN.
I.150	B-ISDN ATM functional characteristics.
I.211	B-ISDN Service aspects.
I.311	B-ISDN General network aspects.
I.321	B-ISDN Protocol Reference Model.
I.327	B-ISDN network functional architecture.
I.361	ATM layer specification for B-ISDN.
I.362	Adaptation layer functional description for B-ISDN.
I.363	Adaptation layer function specification for B-ISDN.
I.413	B-ISDN user–network interfaces.
I.432	B-ISDN UNI–PMD layer specification.
I.610	B-ISDN OAM principles.

QUESTIONS

1 Two cities are 3000 miles (4800 km) apart and are connected by optical fibres and geostationary satellite links. If a 2 × 64 kbits link is set up with one 64 kbits channel being carried on optical fibres and the other by satellite, how many bits of buffering are required to equalize the delay in the two channels? Geostationary satellites are positioned at 22 500miles (36 600 km) above the equator and the speed of electromagnetic waves in free space is 186 000 miless (300 000 kms).

2 Why is the path overhead not associated with the other overheads in the SDH frame structure?

3 Why could the physical interface described in Section 5.1 not simply be operated at 150 Mbits to provide a multipoint broadband interface?

Appendix A

SDL

Kevin Woollard

A.1 INTRODUCTION TO SDL

This book describes the operations and functions of the Integrated Services Digital Network or ISDN. The Standards for this network have resulted from many years of hard work by many national and international bodies. The CCITT are responsible for publishing the I series Recommendations which are the definitive Recommendations on the ISDN. No book on this subject would be complete without a guide to the specification techniques adopted by the CCITT In their work.

The CCITT I series Recommendations, and parts of the network Signalling System No 7, are specified in both natural language (English) and the Specification and Description Language (SDL). In order to understand these Recommendations it is necessary to interpret both parts of the document as they are complementary and additive, neither being sufficient for a complete understanding. This appendix gives an overview of SDL and in particular its use in the ISDN Recommendations.

SDL is one of three specification languages, or Formal Description Techniques (FDTs), supported by the CCITT for specifying networks, services and protocols. It is published in four volumes as Z.100 and annexes. One of the most appealing features of SDL, and certainly one that has provided it with a large body of support, is its Graphical Representation (SDL.GR) which is easy to grasp and allows a rapid understanding of specifications. This ease of use does not mean that SDL should be regarded as a purely graphical notation because it also supports an interchangeable text form or Phrase Representation (SDL.PR) which is machine processable. Even more importantly it is now supported by both formal syntax and semantics which assign a precise and unambiguous meanings to SDL constructs.

The descriptions of SDL constructs given in this appendix are a partial explanation sufficient to provide a general understanding. Furthermore, there are a number of SDL constructs, intended for serious users of the language, that are not mentioned at all. For a full explanation of the language the reader's attention is drawn to the material suggested at the end of the appendix.

A.2 SDL SPECIFICATIONS

During the discussions in the rest of this appendix *italics* are used when ever an SDL term is applied. This is to aid the reader in identifying the SDL terminology and should make it clear when such a term is being applied.

SDL is a rich language capable of defining systems in a structured and hierarchical way. At the top level it divides a *system* into a number of identifiable and interacting *processes*. At this level the *systems* communication paths and their allowed *signals* are specified. At the lower level more detail is given about the operation of the individual *processes* which define the behaviour of the *system*. It is often the case that *process* definitions are used in isolation from the *system* structural parts and an example of such use is the ISDN Recommendations.

The ISDN Recommendations describe both Layer 2 and Layer 3 as single *processes*. The behaviour described details how the protocol operates over the standard ISDN interface. The *process* is a model implementation that may receive *signals* and respond as appropriate. For example the Layer 2 protocol defines a procedure for beginning a communication session which is captured in the SDL. To initialize the protocol it is necessary for an exchange of messages to take place, i.e. the exchange of an 'SABME' and a 'UA'. This is modelled easily in SDL with the 'SABME' being *input* to the *process* and the resultant operation *outputing* a 'UA'. The *process* will then move to a new *state* where the next behaviour is given.

To enhance the robustness of the protocol safeguards are defined, such as limiting the response time to messages and counting the retransmissions of unacknowledged messages. These safeguards are captured in SDL as timers and counters and the result of exceeding either is captured in the SDL *process* description.

The *process* descriptions used for Layers 2 and 3 represent one end of the ISDN interface. A full protocol behaviour is given by considering two such processes interworking with each other, their interconnection being the real world interface of the ISDN.

A.2.1 System structure

This section will introduce the structural parts of SDL which provide the *system* level information of a specification.

Environment

The *environment* is all that is outside the SDL *system* and represents the unspecified parts of the real world. The SDL *system* and the *environment* communicate with *signals* pased over *channels*. The behaviour of the *environment* is undefined; that is to say, we must take the world as it comes and not how we might like it to react.

System

The *system* defines the boundaries of the SDL specification along with the points of interface to the *environment*. It also shows some of the hierarchy of the specification in terms of a number of *blocks* and their intercommunications via *channels*. No behaviour is described at this level. A *system* diagram shows the *system*, its *blocks* and their communication as shown in Figure A.1.

SYSTEM SDLsyntax

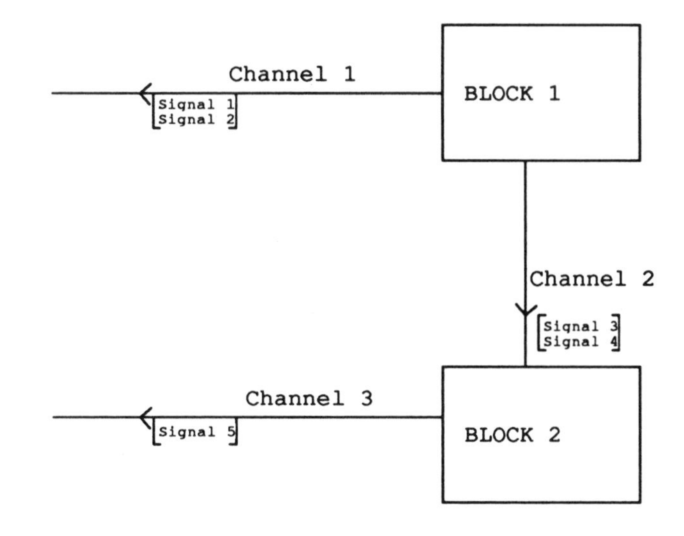

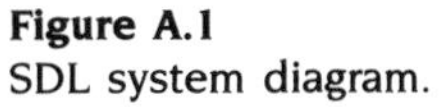

Figure A.1
SDL system diagram.

Block

The *block* is the first level of ordering in the *system* and each *block* represents a functional unit. A *block* can communicate with other *blocks* or with the *environment* using *signals* transmitted over *channels*. A *block* can be subdivided into *blocks*, if further hierarchy is required, or *processes* if a behaviour description is appropriate. *Processes* and *blocks* may not be mixed within a *block* and a *block* subdivided into *blocks* will show the *channels* and *blocks* in

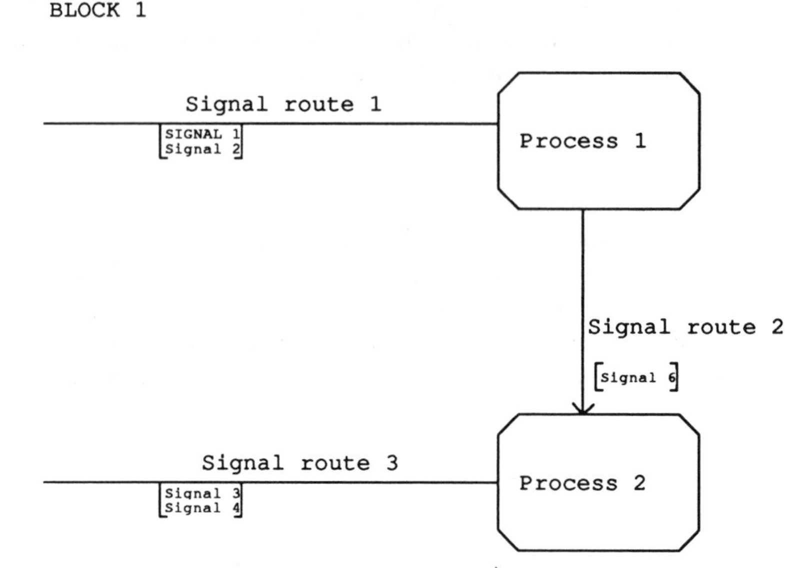

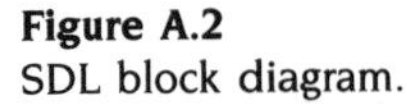

Figure A.2
SDL block diagram.

the same way as the *system* diagram. A *block* that comprises *processes* will show a number of *processes* and their communication paths and *signals*. The communication between *processes* is via *signal routes* as is communication between a *process* and the *block* boundary. No behaviour is described at the *block* level and each *block* has an associated *block* diagram as shown in Figure A.2.

Process

Each SDL *process* is a *state* transition system modelled as an Extended Finite State Machine (EFSM) which allows the specification of behaviour and its interaction with data. It is the use of data that 'extends' the state machine model from a Finite State Machine (FSM), which can only describe behaviour in terms of inputs and outputs. The data can be input as *parameters* or exist in the *system* as *variables*. To the user this model provides a flexible notation that embodies the concepts of *states, inputs, outputs, tasks, timers* and *decisions*.

An SDL *process* will wait in a *state* until it receives an *input*. Then it will move out of its current *state* and perform a number of operations, its last operation will be to move into a new *state*. This action of moving between *states* and performing operations is termed a *transition*. The operations could be to transmit one or more *outputs*, execute a number of *tasks, set* or *reset* timers and make a number of *decisions*. Depending on the *decisions* the *process* will move on to another *state*. Of course the *transition* executed is dependant on the *signal* received by the *process* as different *inputs* will lead to different *transitions*.

A.2.2. SDL process structure

A simple explanation of the operation of a *process* is given in the section above along with an indication of some of the constructs. With the use of the above concepts the specifier can develop complex *systems* that communicate with their *environment* through the use of *signals*, and control their data through the use of *tasks*. The exact behaviour of these *systems* is itself dynamic and dependant on the results of *decisions*. To take our understanding of the notation further it is necessary to introduce the various elements of the language and describe them on a construct by construct basis. The components of the *process* are described below and their graphical representation is given in Figure A.3.

Start

The *start*, as its name suggests, is the initial starting point of the SDL *process*. Once a *process* has been initialized it will move immediately out of the *start* and execute the associated *transition*. The *start* is a special case of a *state* as it does not require an *input* but rather executes a *transition* spontaneously. During this initial *transition* is is possible for the *system* to execute *tasks*, *decisions*, *timers* and *outputs* as in any other *transition* of the *system*.

State

The *state* is the natural stable condition of the *process* and a *process* is defined with a number of possible *states*. A *process* can change *state* by way of a *transition* which is triggered by the *input* of a *signal* from another *process*, or from outside the *system* (the *environment*). The *state* will identify a number of *inputs* that cause it to execute a *transition*, during which a number of operations are carried out. Once a *transition* has occurred the *process* enters a new *state* which may be any of the valid *states* declared for the *process*, including the previous *state*.

Those *signals* for which a corresponding *input* is shown in the current *state* of the *process* are said to be explicitly consumed. *Signals* not shown in a particular *state* are ignored by the process and are said to be implicitly

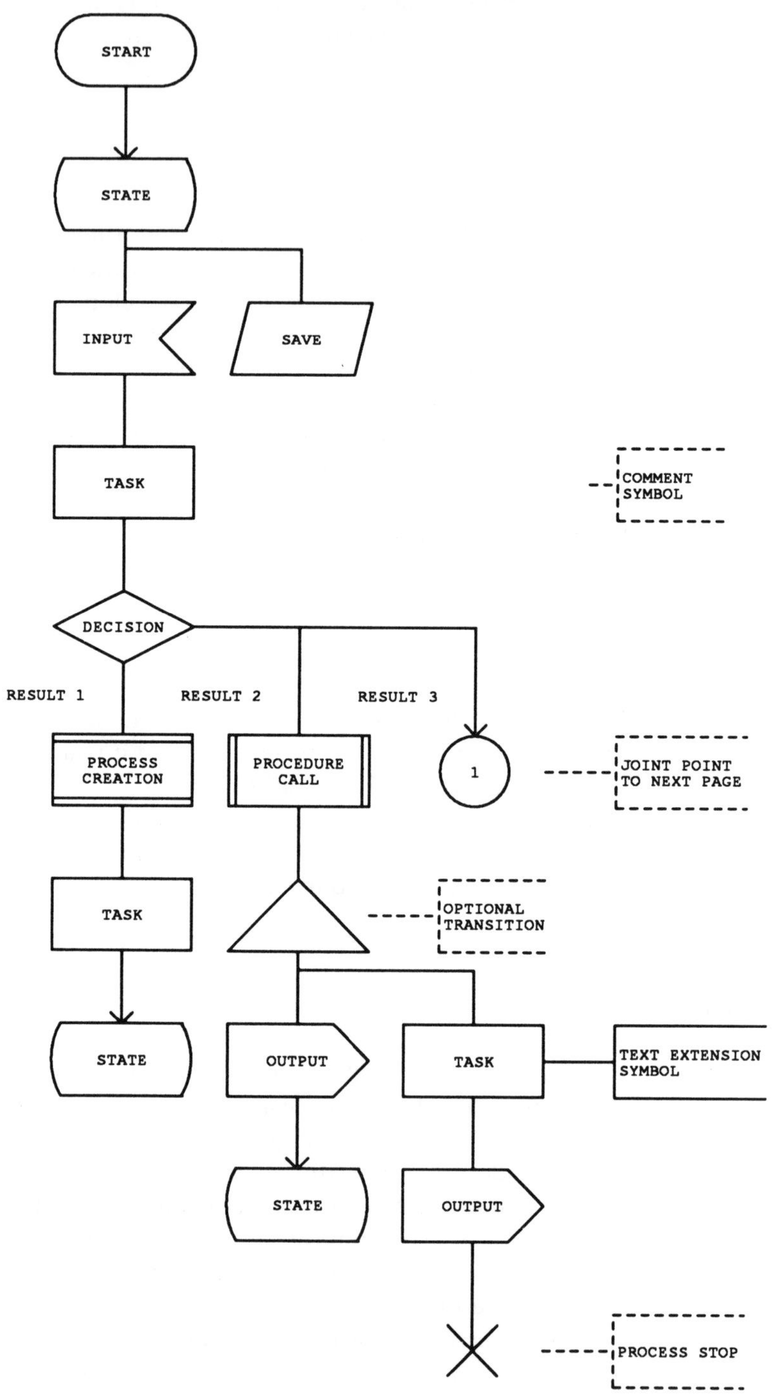

Figure A.3
Constructs used in SDL process diagrams.

consumed. In such a case the *signal* is considered to be consumed, no *transition* undertaken and the *process* remains in the same *state*.

Input

A lot has been said about *inputs* in the section above, which is not surprising due to their close association with *states*. An *input* is supplied to the *process* in the form of a *signal* from either another *process* in the *system* or from the *environment*. Because the communication in the *system* is asynchronous, i.e. independent of the current condition of the *processes*, the *signal* may arrive during a *transition*. Under such circumstances the *process* is unable to consume the *signal* and so there is a need for a queueing mechanism. SDL uses the concept of FIFO (First In First Out) queues where all signals arriving at a *process* are queued in order of arrival and removed from the queue in the same order.

Once a *signal* has been queued the *process* is able to consume it as an *input* to its next *state*. If there is no defined behaviour for that *signal* the implicit consumption will discard it from the queue allowing the next *signal* to be considered. This will continue until either all the *signals* in the queue are discarded or, *input* to the *process* resulting in a *transition*. Either way all the *signals* will be removed from the queue and the *process* will wait in a *state* until the next *signal* arrives.

Signals may carry *parameters*, which can be assigned to *variables* in the *input* expression, and used in subsequent operations such as *tasks* and *decisions*. There can be any number of *parameters* associated with a *signal* but only those assigned to *variables* in the *input* expression are available for further manipulation.

Save

If a *state* cannot deal with a *signal* but the *signal* is too important to be discarded it may be *saved* for later processing. A *signal* queued for the *process* but shown to be *saved* in the current *state* is not removed from the FIFO queue but is left and the second, or next non-*saved signal* is consumed. After executing a *transition* the *process* may enter a different *state* and the *signals* previously *saved* are available for consumption in the normal way. It is possible that *signals* can be *saved* in successive *states*.

Task

The *task* allows the process to perform operations which control some particular aspect of the *system*. The expressions contained in a *task* may be either data manipulations, informal textual instructions or the control of timers.

The data manipulations allow counters to be incremented, *variables* evaluated, or assigned to other *variables*, and structures and strings to be constructed.

Comments

To allow the specifier to explain his work the *comment* symbol may be associated with any of the other elements of the *process*.

Informal text allows the specifier to give information about the *system* behaviour that is not easily captured in data. Such information may relate to behaviour not under the control of the specification being developed but relevant to its operation. In these conditions a textual note to modify or retrieve information is wholly sufficient for the specifier's purpose.

Decision

To allow some dynamic control in the behaviour of an SDL specification the language supports the concept of *decisions*. It is the result of evaluating the expression that leads to the alternative behaviour of the specification. For each possible result that can be generated from the expression there is an associated *transition* in the specification.

A simple example would be the comparison of a *variable* with a *constant*, e.g. $a > 10$. In this case the value of the *variable* a is compared to the value 10. If a is 10 or less the result 'false' is returned, if a is greater than 10 the result *'true'* is returned. Associated with such a *decision* will be two branches, one associated with each of the potential results, 'true' and 'false'.

The contents of the *decision* may be data operations, as described above, or informal text.

Process creation

SDL supports the concept of *process creation* allowing the specifier to generate extra *processes* in the *system* as required. An example of such a requirement might be the use of separate *processes* to handle successive communication sessions. In this way the *processes* are generated as required and removed from the *system* as the requirement is exhausted allowing the *system* to remain in its simplest form.

Processes created in this way have their behaviour defined in the same way as other *processes* in the *system* and any number of concurrent instances of a *process* may be allowed.

Procedure call

To allow some compression in its notation, SDL allows common behaviour to be defined in a *procedure*. The scope of a *procedure* is limited to its associated *process* and may be called from any point in that *process*. The body of a *procedure* contains the same constructs as the body of a *process* including *states, inputs, outputs, decisions, tasks*, etc. In this way complicated behaviours can be specified in a *procedure*. The *inputs* to a *procedure* are taken from the FIFO queue associated with the controlling *process*. The constructs of a *procedure* are similar to those of the *process* but with a different *start* symbol and the addition of a *procedure return*.

Once a *procedure* has been called, control is passed from the *process* and remains with the *procedure* until a *return* is encounted. After a *return* the control is passed back to the *process* at the point the call was made.

Optional transition

To allow some flexibility at implementation time the *optional transition* construct may be used. This identifies two or more possible *transitions* or part *transitions* that can be selected by the implementer. This construct is often used if a CCITT Recommendation has to support different national Standards allowing the differences to be identified as optional.

Text extensions

Because of the limited size of the pictorial elements of the SDL graphical form it is necessary to allow text to overflow into a *text extension* symbol. All text in such a symbol must be considered part of the construct with which it is associated.

Output

The *output* is part of the communication of the *process* with its surrounding *system* and *environment*. An *output* may occur any time during a *transition* and any number of *outputs* are allowed. *Outputs* are delivered to the appropriate destinations as *signals* and may carry data.

Timers

Timers are used to specify the time dependent parts of the system. For example, in the ISDN error recovery *procedures* will be started if a certain time elapses between frames. *Timers* may be *set* or *reset* causing the starting and stopping of *timers* respectively. Setting a *timer* activates a timing process which is responsible for the execution of timers. Once *set*, a *timer* will run for a predetermined period at the end of which an *output* will be sent to the originating *process*. A running timer can be stopped with *reset* or restarted with *set*.

Given the elements of a *process* described above it is possible to construct very useful specifications and indeed many of the ISDN specifications use no more than these. This subset of SDL provides a good understanding of the notation and should allow the reader to interpret the CCITT Blue Books on ISDN.

A.2.3. Communcation

Within the SDL environment there is a model of communication based on asynchronous connections and queues. Within this model there are two specific communication media identified with different transmission characteristics. It is left to the discretion of the specifier to construct his or her system to take advantage of the most appropriate medium.

Channel

A *channel* will only transmit *signals* specified as part of its *signal list;* no other message can be suported. *Signal* order will be maintained within the *channel*, i.e. *signals* will be output in the same order as they were input, and lastly a *channel* may subject *signals* to unspecified delays. *Channels* originate and terminate on the boundaries of the *system* and/or *block*. At a *block* boundary the *channel* is associated with one or more *channels* or *signal routes* which in turn connect to sub-*blocks* or *processes*.

Signal route

Signal routes are similar to *channels* with *signal* order being maintained. They differ, however, because *signal routes* insert no delay. *Signal routes* originate and terminate on the boundaries to *processes* or the *block* containing the processes. Only *signals* defined as part of the *signal route's signal list* can be supported.

A.2.4. Data in SDL

It is possible to introduce data into SDL specifications in two ways. The simplest method is to treat all data as informal text allowing the specifier to use any data in any way he wishes. The problem with this approach is that his intentions are not always clear. For example, the specification may require some operation involving queue manipulation. If the explanation of the operations is informal, or only implied by usage, then it is highly probably that different people will interpret different meanings. This problem of informally defined specifications can be resolved by introducing more rigour into the data definitions. In SDL the ability to formally define data is provided through the use of the Abstract Data Typing (ADT) part of the language.

Abstract data types allow a complete and mathematically based definition of a data type including its *literals* (or alphabet), its *operators*, and the *axioms* relating the *operators* to the *literals*.

The *literals* of an ADT define the minimum set of elements from which all other elements of the same type can be constructed. For example, the *literals* of the data type integer are all the whole numbers from 0 to 9. Given this set all the whole numbers from minus infinity to plus infinity can be constructed. For example, the integer 392 is a sequence of 3, 9 and 2 each of which is a *literal* of the type integer.

The *operators* of an ADT are all the *operations* that can be performed on the *literals*. For example, in the integer data type *operators* include +, −, =, etc. These *operators* are defined in terms of their functionality; for example, the *operator* + takes two *literals* of the *type* integer and returns a single *literal* of the *type* integer. Of course this definition says nothing about the rules of addition but merely identifies the function of the *operator*.

The *axioms* of the ADT provide all the rules needed to apply the *operators* to the *literals*. It may require a number of *axioms* to define each *operator* under various circumstances. For example, the rules for subtraction are:

$$a-a \quad == 0$$
$$(a-b)-c == a-(b+c)$$
$$(a-b)+c == (a+c)-b$$
$$a-(b-c) == (a+c)-b$$

where
a,b,c are variables representing any integer;

+ is the addition *operator* which has already been defined in the definition of integer;
== is part of the data typing language and means 'is defined as being equal to';
() is part of the data typing language and indicates priority.

Given this definition of subtraction, any expression containing the *operator* '−' can be evaluated.

Such a set of *axioms* will exist for all *operators* in the data type.

In general it is up to the specifier to define the data that his specification uses. There is no restriction on the data types that can be defined. However, as the task of specifying data types is an overhead for the specifier, SDL supports a small library of predefined types that are available for use without further definition. The types supported are: boolean, character, string generator, charstring, integer, natural, real, array generator, powerset generator, process identity, duration, and time.

These types are sufficient for many applications and a full description of each is given in Z.100.

A.3. AN EXAMPLE OF SDL APPLIED TO SIGNALLING SYSTEMS

As an example of an SDL specification Figures A.4, A.5, and A.6 show part of a specification for Layer 2 of the Message Transfer Part (MTP) of CCITT signalling system No. 7. The complete specification is complex and large and only a small subset is presented in the figures.

Figure A.4 shows the *system* level diagram which details the interfaces between the *system* being specified and the *environment*. In this case there are two *channels* into the *system* and two *channels* out, each with its own associated list of supported *signals*. Also presented at the *system* level are the abstract data definitions; in the example there are four *newtypes*, which will be used later in the specification. The *system* described in this specification has only one *block* which is expanded in the *block* diagram given in Figure A5.

In the *block* diagram there are two *processes* and a number of *signal routes* connecting the *processes* to each other or to the *block* boundary. As with the *channels* in the previous diagram, each *signal route* has an associated list of supported *signals*. As this is only a part of the full specification there are no signals detailed for one of the processes. By inspection of the *system*

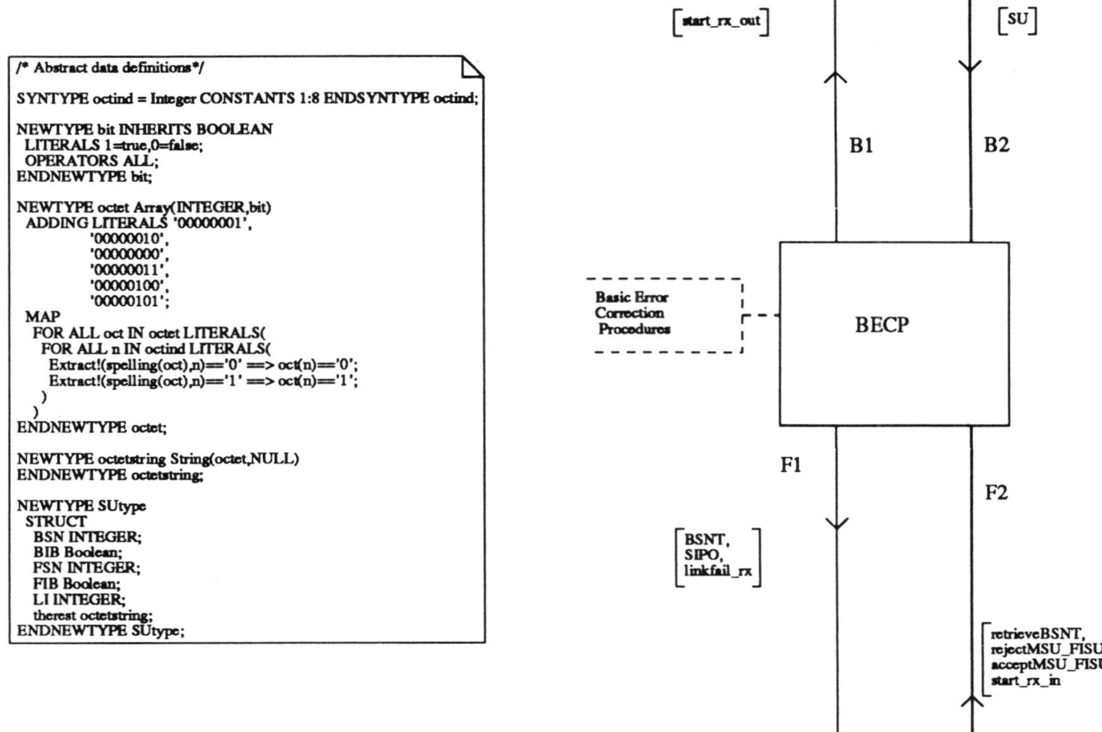

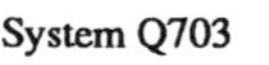

Figure A.4
System definition for part of CCITT signalling system No 7.

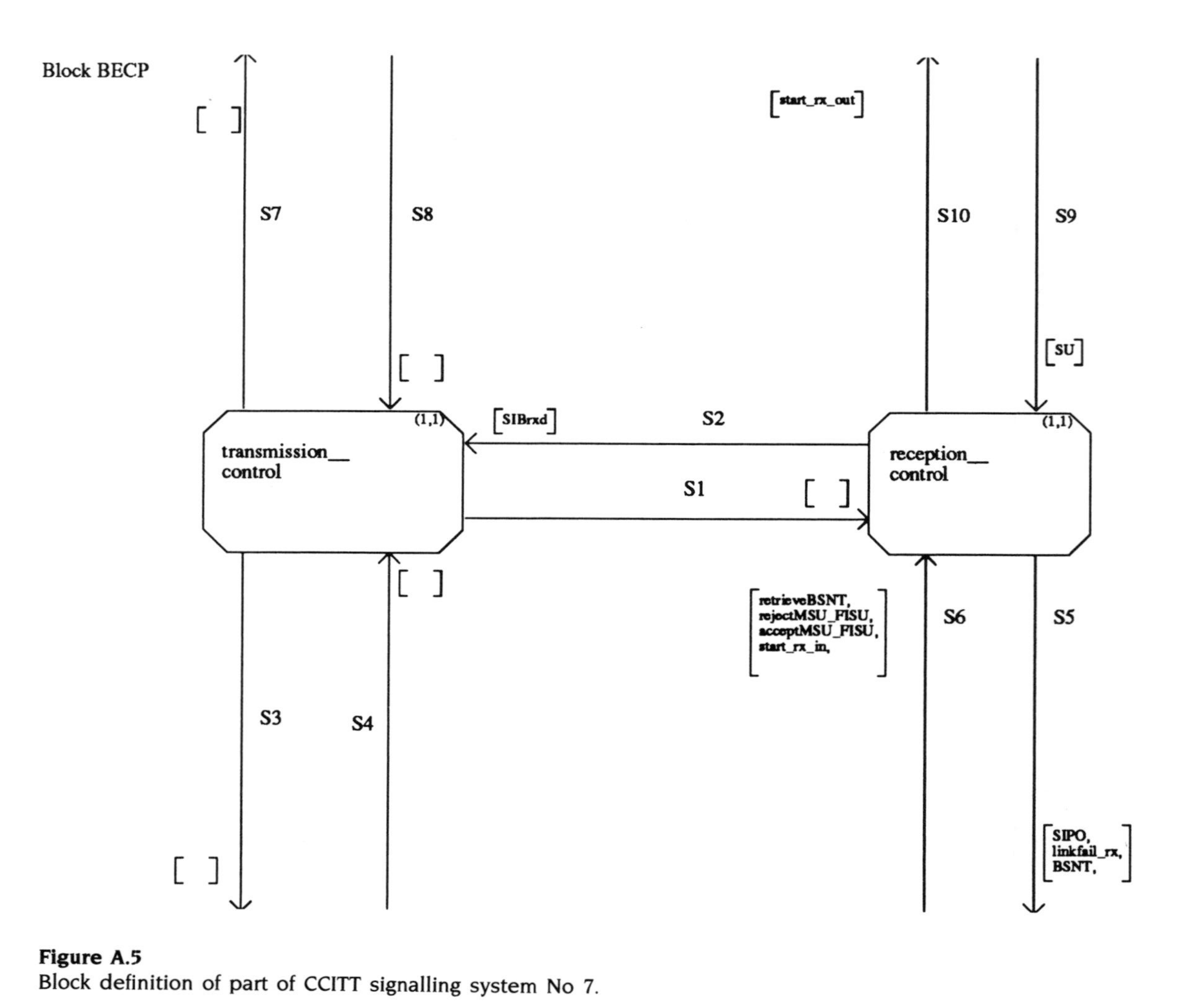

Figure A.5
Block definition of part of CCITT signalling system No 7.

diagram and the *block* diagram a mapping can be interpreted between the *channels* and the *signal routes* based on the *signals* supported. It is possible for a number of *signal routes* to connect to the same *channel* as required by this example.

Part of the behaviour of one of the *processes* in the *block* is given in Figure A.6. In this figure we see the *process* description and the data declaration associated with it.

After the *start* of the *system* the *process* moves to its first *state* where it is able to consume one of three *inputs*. These *inputs* are previously declared as *signals* in the *system* and *block* diagrams and so their origin is known to be the *environment*. Having consumed the *inputs* the *process* will execute the associated *transition*.

If the *input* 'SU (signalunit)' is consumed the *process* moves to evaluate a *decision*. In this case the *decision* is based on an element of the structured local *variable* 'signalunit' which was assigned the value of the *parameter* in the *input* 'SU'. If the *parameter* returns the value for 'LI' of 1 or 2, the *decision* returns the result 'true' and the *transition* continues down the appropriate branch. This leads the *process* to another *decision* based on the same *parameter* but a different element, i.e. the element named 'therest'. (The complete structure of the *parameter* carried by 'SU' is given in the abstract data typing of '*newtype* SUtype', given in the *system* level diagram.) Three results are possible from this second *decision*, two producing *outputs* and one returning the *process* directly to its previous *state*. Of the two *outputs* possible one is passed to the *environment* and one is passed to the other *process* in the *system* for further processing. It is interesting to note that the internally passed *signal* (SIBrxd) is never passed out of the *block* diagram and consequently can not be seen in the *system* diagram.

Had the result of the first *decision* in the *transition* been 'false' the behaviour of the *process* would have been significantly different, involving *tasks*, *decisions*, *outputs* and possibly behaviour not shown in Figure A.6 associated with the *join point*.

The example given in this annex is only a small part of the full specification; however, it is clear to see that the use of SDL allows a clear and precise specification of message based systems.

Due to the *state* based model it is relatively easy to move from a specification to an implementation, making SDL a powerful tool for all stages in the protocol development life cycle.

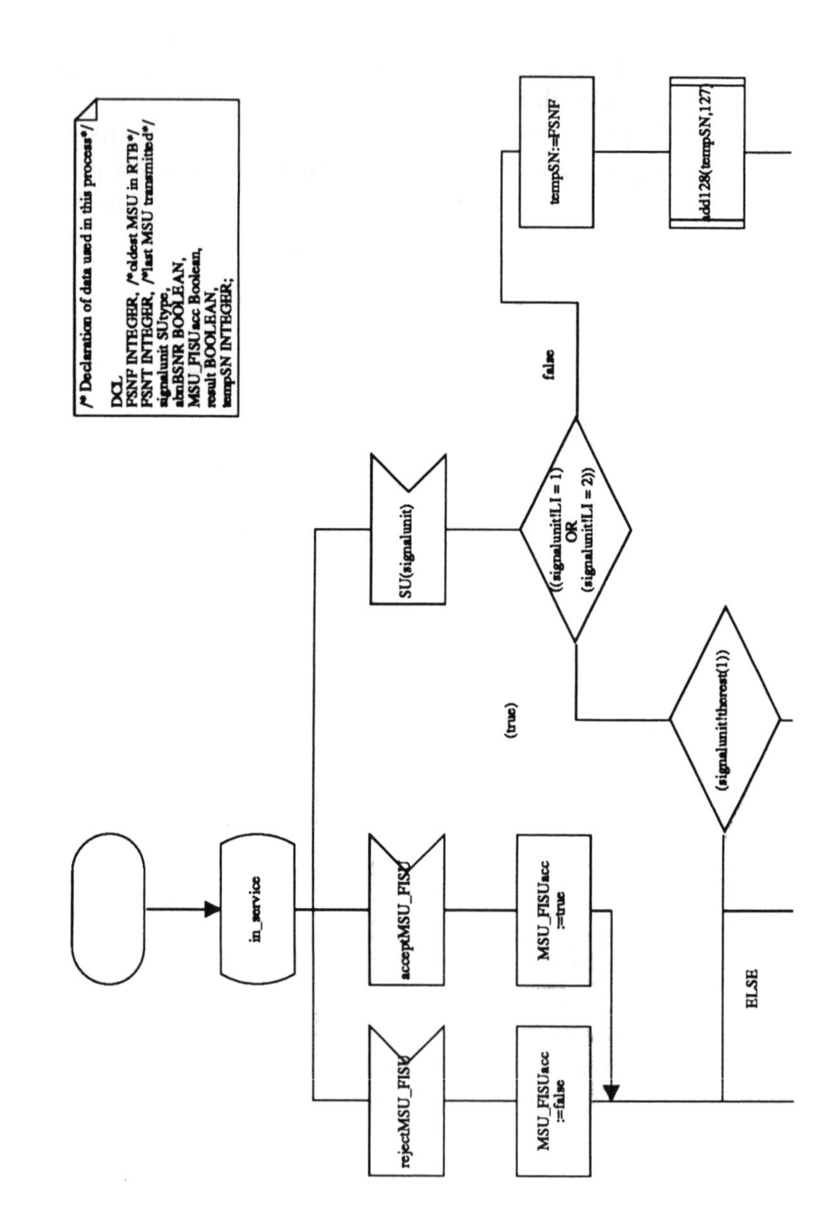
/* Declaration of data used in this process*/
DCL
FSNF INTEGER, /*oldest MSU in RTB*/
FSNT INTEGER, /*last MSU transmitted*/
signalunit SUtype,
abnBSNR BOOLEAN,
MSU_FISUacc Boolean,
result BOOLEAN,
tempSN INTEGER;
in_service
acceptMSU_FISU
rejectMSU_FISU
MSU_FISUacc
:=true
MSU_FISUacc
:=false
SU(signalunit)
((signalunit!LI = 1)
OR
(signalunit!LI = 2))
false
(true)
tempSN:=FSNF
add128(tempSN,127)
(signalunit!therest(1))
ELSE

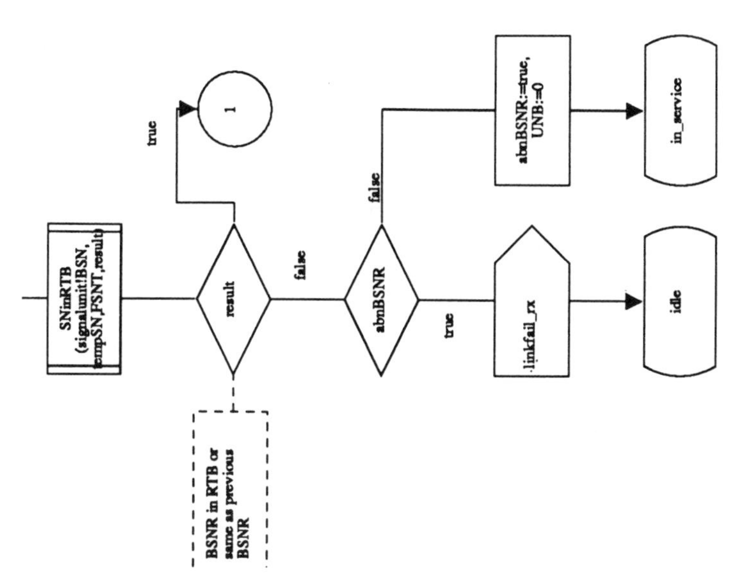

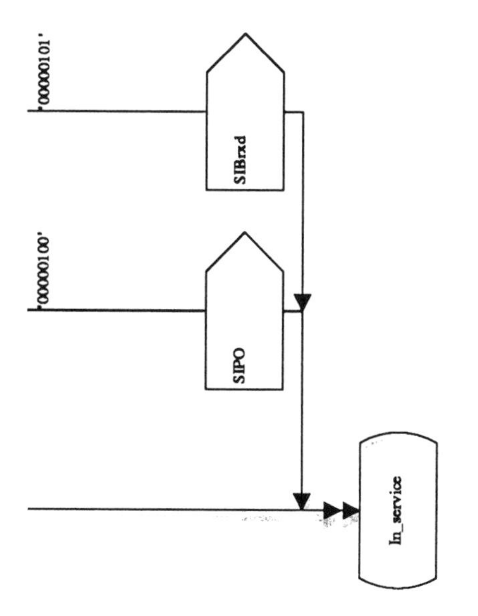

Figure A.6
Taken from behaviour desription of CCITT signalling system No 7.

REFERENCES

CCITT Recommendation Z.100 and annexes A, B, C, E, F1, F2 and F3.

R. Saracco, J.R.W. Smith and R. Reed, *Telecommunications Systems Engineering using SDL*, 1989 North-Holland.

F. Belina, D. Hogrefe and A. Sarma, *SDL: with Applications from Protocol Specification*, 1991 Prentice Hall.

K. Turner (ed), *Using Formed Description Techniques: an Introduction to Estelle, LOTOS and SDL*, 1992 John Wiley & Sons Ltd (to be published).

Appendix B

Answers to Questions

Chapter 1

1 What are the advantages of digital transmission over analogue transmission? In view of these advantages why did the analogue telephone network predominate for 100 years?

Digital transmission has the advantage that the signal may be regenerated, and, providing this is done before distortion and interference reaches the point where digits become unrecognizable, no errors will result and hence there will be no impairment to the audio signal. Telephony predominated because speech proved much more attractive to the human user in comparison with alternatives such as teleprinter or morse code. The technique to convert speech into digital form was not conceived until 1937, and could not be realized in an economical form until transistors became widely available in the 1960s.

2 If the signal power equals the noise power in a channel of bandwidth 1 Hz, what is the theoretical information rate in bit/s which can be carried through this channel.

Shannon's theorem: $I = F \log_2 (1 + S/N)$. If $F = 1$, and $S = N$ then $I = 1$ bit/s.

3 A telephone, data terminal (with modem) and facsimile machine can all be connected in parallel on an existing analogue public switched telephone network. What advantages would be obtained by connecting:
(a) the non-voice terminals to a data network?
(b) all the terminals to the ISDN?

(a) The data terminal may not need a modem, depending on the form of the access. Call set-up will be much more rapid than would be provided by early telephone networks. The data transfer rate may well be higher and error rates lower. On packet networks simultaneous virtual connections to several destinations can be supported. Group 4 fax machines could operate at 64 kbit/s if the network can operate at that rate.
(b) All terminals can simply be connected in parallel. Data and Group 4 fax machines can operate at 64 kbit/s. Incoming calls will automatically be routed to the appropriate terminal and two channels are available on a basic rate access.

4 Two users make arrangements to exchange information. Initially they use a data terminal via a modem on the public switched telephone network. They then repeat the same exercise via a packet network. Which of the ISO 7 layers are the same in both cases?

The 7-layer model can only be considered in relation to a particular point in the network. Here we consider the interface between the terminal and the modem or network.

At Layer 1 the interface may well be the same both for access to a packet network or the telephone network, typically V24 or RS232. If a dataline is used for packet access then Layer 1 may be X.21.

Layer 2 for packet access is HDLC. This may well also be used on the telephone network for error control.

Layer 3 will contain the biggest differences in the call set-up phase. The packet network set-up is defined in X.25 whilst to set up a call on the telephone network will either involve the use of a proprietary protocol (such as Hayes) from the terminal to the modem which will then generate dial pulses or MF tones, or a separate dial or MF keypad can be used. In the data transfer phase the packet network uses X.25, and a subset of this could be used on the telephone network for flow control and with the option of providing multiplexing.

The higher layers will be the same in both cases.

Chapter 2

1 Voice frequencies are limited to 3.4 kHz. Give two reasons why a PCM sampling rate of 6.8 kHz is not used.

1 Filters do not have an infinitely fast rate of cut-off.
2 Leakage of the sampling waveform into FDM systems would result in interfering tones unless the sampling waveform is at a multiple of 4 kHz, so that the interfering tones would be demodulated by the FDM system as low frequencies which are inaudible.

2 Music requires a bandwidth of 16 kHz. Experience has shown that the simple compression technique described for speech is not satisfactory. If it were decided to use linear coding with an overall coding accuracy of 0.03% of the peak amplitude, what overall bit rate is required?

0.03% represents 1 part in 3333. 12 bits offers a precision of 1 part in 4096, and a further bit is required to indicate polarity. A minimum sampling rate of $2 \times 16 = 32$ kHz is required. Hence a bit rate of $32 \times 13 = 416$ kbit/s is required.

3 Why is it not possible for an A-law or μ-law PCM encoder to continuously generate a long string of 0s on the transmission system?

The idle coded signal could be zero volts, represented by continuous 0s. To avoid this A-law encoders invert alternate bits, and μ-law encoders invert all bits. An all-0s code word can still be generated but as there is no DC connection to the coder, and the all-0s code word no longer represents 0 volts, then the next code word must be different.

Chapter 3

1 What percentage of transmitted bits are allocated to signalling on 24 and 30 channel PCM systems?

24 channel system: $[(24/193)/6] \times 100 = 2.07\%$ in the channel associated signalling format. In the common channel 23 channel form it is $(8/193)/100 = 4.1\%$
30 channel system: $(8/256) \times 100 = 3.13\%$ in the common channel form. For

channel associated signalling systems, in which a multiframe structure is provided and the signalling bits relating to channels 0 and 16 obviously do not carry signalling, the correct answer is $3.13 \times 30/32 = 2.9\%$.

2 A sequence of binary digits
1000010000001100000001... is encoded in HDB3 as +000+− .... Complete the coding. How would the binary sequence be encoded in B8ZS?

In HDB3: + 0 0 0 + − 0 0 0 − 0 + − + 0 0 + − 0 0 − + . . .
In B8ZS: + 0 0 0 0 − 0 0 0 0 0 + − 0 0 + 0 + − 0 − + . . . (or the inverse)

3 Time switches are much less costly to build than space switches because of the availability of VLSI. Why then are space switches incorporated into switching nodes?

The space switch in a time division multiplexed system behaves as if it were N space switches, where N is the number of channels multiplexed together; hence its cost per channel may be divided by N and so is not of great consequence. The space switch provides routes to different physical destinations.

4 The local exchange consists of three main parts - a concentrator, route switch and processor. What are the merits of locating them all in the same place, and what are the merits of separating them?

The processor may be separated from the other components to give economies of scale (by having the processor covering a large area), simplifying security and software maintenance and upgrades. The concentrator may be separated to take it nearer to the customer to reduce access line lengths. The disadvantage of the separation lies in the potential for failure of the lines linking the parts together.

Chapter 4

1 The transmission system chosen for the local network is designed to perform adequately on about 99% of local network pairs. If satisfactory performance were required on only 50% of pairs how many ISDN B channels could be carried on each pair?

From Figure 4.3 it can be seen that 99% of pairs have loop resistances of roughly 1200 Ω or less. It can be seen that the corresponding figure for 50% of pairs is about 650 Ω (USA) or 400 Ω (UK). Looking at figure 4.13 it can be seen that a 1200 Ω loop has a theoretical capacity of 600 kbit/s; the practical carrying capacity is 144 kbit/s indicating that about $144/600 = 0.24$ of the theoretical rate is practical. From Figure 4.13 it can be seen that 650 Ω and 400 Ω loops have a theoretical capacities of 3000 and (with some extrapolation) 10 000 kbit/s respectively. Converting to realizable rates by multiplying by 0.24 gives 720 and 2400 kbit/s respectively, suggesting that 11 B channels could be carried in the USA and 37 in the UK.

2 Why is a simple hybrid arrangement inadequate for digital duplex transmission in the local network?

Hybrid networks depend for their operation on being able to create a balance network of identical impedance to the cable pair. With the bandwidths involved and the varied nature of the pairs used in the local network this is not achievable with sufficient precision.

3 Echo cancellers have two basic structures—look-up table and transversal filter. What are the advantages and disadvantages of each?

The look-up table is in principle very simple; it is rugged as it does not assume any linearity in the system. On the other hand its size increases exponentially with the echo length and it converges slowly. The transversal filter only requires as many elements as there are samples in the echo, and converges rapidly; however, its operation does rely on the output of the transmitter being linear.

4 Why are scramblers used on local network transmission systems?

Echo cancellers operate by eliminating any correlation between the received signal and the transmitted signal. If there is some inherent correlation, for example due to similar synchronizing patterns in both directions of transmission, then the echo canceller will not converge correctly. To avoid this each direction of transmission is scrambled using a different scrambler.

Chapter 5

1 How many pins are there on the ISDN connector? What are they used for?

There are eight pins. Two transmit from NT1 to terminal (pins 4 and 5); two transmit from terminal to NT1 (3 and 6); two provide a power feed from terminal to NT1, if needed (1 and 2); two provide a power feed from NT1 to terminal (7 and 8).

2 The incoming information rate from the exchange is 144 kbit/s but the ISDN interface operates at 192 kbit/s. How are the extra bits used?

Framing, DC balancing, echo channel, activation and multiframing. There is also a spare bit for possible future use.

3 Starting from an idle circuit, how is a simpe telephony call initiated including the operation of all layers?

The initiating telephone will send INFO 1 signal to the NT1 in the D channel.

If the line transmission system is idle then the NT1 will start transmitting to the exchange, and the exchange will also transmit, timing circuits will have to align themselves and echo cancellers and equalizers will have to converge to their correct settings.

The NT1 will then return INFO 2 to the terminal. When terminal and NT1 are synchronized then the standard (INFO 3 and 4) format will be exchanged at Layer 1 on the bus.

D-channel Layer 2 will now establish a LAP by sending from the terminal SABME which will be acknowledged by UA. Layer 2 can now carry Layer 3 information from terminal to the exchange. The terminal will then send a SETUP message to the exchange, which will acknowledge it. The terminal will send INFO messages containing the destination number. When these are complete the exchange will return to the terminal CALL PROCEEDING.

At the destination exchange the line transmission system will be activated and the NT1 will send INFO 2 to activate all the terminals. Layer 2 will also be established between the destination terminal and the destination exchange.

Layer 3 from the destination exchange will send a SETUP message to the destination terminal, causing a bell, or other device, to sound and returning an alerting signal to the originating terminal.

When the destination terminal is answered a CONNECT message is sent from the destination terminal to the destination exchange. It acknowledges it to the destination terminal and also passes it back to the originating exchange, which passes it to the originating terminal, which acknowledges it to its exchange.

Conversation can now start.

4 What is the difference between stimulus and functional signalling and what are the advantages of each?

With stimulus signalling the terminal does not have any knowledge of the service being invoked. The terminal simply tells the network that certain buttons have been pressed, usually including * and #, and the network interprets this. This procedure is simple to arrange, requiring no collaboration between terminal manufacturer and network provider, but is complicated for the user when several calls are involved (e.g. conference calls). Functional signalling requires both terminal and network to carry records of the state of the service.

Chapter 6

1 The European and North American network digital transmission systems carry 30 and 24 channels respectively. However, the primary rate interfaces offered to the customer, whilst remaining at 30 channels in Europe, drop to 23 channels in North America. Why the difference?

The 30 channel format had 64 kbit/s allocated for signalling. The 24 channel format provided signalling by stealing from the 64 kbit/s traffic channels; hence to provide clear 64 kbit/s channels the signalling had to be accommodated elsewhere and the 24th channel is used.

2 Why is it desirable to have compatibility between user-network signalling and inter-PABX signalling, and what are the different needs?

Compatibility is desirable to allow mixing of private and public network traffic on a single primary rate multiplex. Compatibility also allows the re-use of some software functions for both types of traffic. Public network signalling is based on the premise that the network controls the peripheral devices, whereas the private network signalling exists between equals and so is symmetrical.

Chapter 7

1 When X.25 was introduced it made a feature of the fact that signalling and data were treated in a common manner. Frame mode services now make a feature of the separation of signalling and end user data. Explain the apparent contradiction.

X.25 offered great flexibility compared with circuit-switched alternatives, mainly due to the inherent multiplexing from the virtual circuit environment. At the time, data capacity requirements were less important. However, it is now appreciated that the need for the network nodes to examine every packet to determine whether it is control or data inevitably reduces potential throughput. Frame mode services enable designers to separate the complex, software-driven control from the simple, fast hardware switch.

2 Packet and frame mode services can operate in both the B channel and the D channel. What are the merits of the B channel compared with the D channel for this purpose?

The B channel offers 64 kbit/s compared to the D channel's 16 kbit/s. It also may be free of some of the frame length limitations imposed by LAPD.

3 Figure 7.1 implies that in some circumstances a network with 'poor' flow control can carry more traffic than one with 'good' flow control. Why is this?

The objective of the flow control is to maintain the integrity of the network so that it operates in a reliable and predictable manner. To do this it may restrict throughput at times to ensure that congestion does not jeopardize traffic elsewhere.

Chapter 8

1 Why is it necessary to define teleservices?

It is not sufficient to simply establish a 64 kbit/s channel. It is also necessary that the data stream has the same significance to each terminal — for example, by specifying the speech or picture coding technique and the format in which the information is transmitted. This definition is a teleservice.

2 In an ADPCM system the decoding depends on the previous information received. How then are errors in transmission tolerated so that permanent errors of decoding do not result?

The quantization is determined not only by the previous signal range, but also by the passage of time to allow it to tend to a fixed value, so any error will eventually disappear.

3 The high band signal in sub-band ADPCM encodes signals up to 7 kHz but is only sampled at 8 kHz. Produce a diagram like the one in the top right-hand corner of Figure 2.1 to show that the resulting sidebands do not overlap.

If the arrow represents signals originating in the range 4→7 kHz then the resulting sidebands will lie:

```
0          8            16             24  kHz
|<-------->|<---------->|<------------>|   ....
```

4 If a Group 4 fax machine is transmitting a page consisting of vertical, constant-width black and white stripes, what digital sequence will be generated to transmit this after the first line (ignore end of lines)? What code sequences will be generated if the stripes are at 45 degrees to left and right?

Vertical stripes will be encoded in vertical mode with value V(0) for each edge. This is encoded as a 1 for each edge. With 45 degree stripes the edge will move by one pel for each line of pels, which is transmitted as 011 or 010 according to the direction.

5 Text displays on a screen may be encoded by the DCT or sent as International Alphabet 5 (ASCII) characters. By comparing the bits per pel required for each method, decide on the relative efficiency of the two for the purpose. Note that both forms of text appear in Figure 8.10—look for the name on the back of the boat.

The letters on the back of the boat become readable at about 0.75 bit/pel. The text in the caption is part of a 24 row by 40 character display, each character of which requires 8 bits to transmit. The whole display therefore requires $24 \times 40 \times 8 = 7680$ bits. The number of pels is 262k (see Figure 8.9). Therefore the bits/pel of this form of transmission would seem to be $7680/262\,000 = 0.03$. However, each letter in the caption is about twice the linear size, hence four times the area, of the letters in the name on the back of the boat. Thus $0.03 \times 4 = 0.12$ bit/pel is a rather fairer figure. Thus the ASCII method of encoding is about $0.75/0.12 = 6$ times more efficient than DCT for transmitting text. This is not surprising since ASCII is designed for this purpose while the DCT can cope with any arbitrary shape.

Chapter 9

1 Two cities are 3000 miles (4800 km) apart and are connected by optical fibres and geostationary satellite links. If a 2×64 kbit/s link is set up with one 64 kbit/s channel being carried on optical fibres and the other by satellite, how many bits of buffering are required to equalize the delay in the two channels? Geostationary satellites are positioned 22 500 miles (36 000 km) above the equator and the speed of electromagnetic waves in free space is 18 600 miles/s (30 000 km/s).

A simple calculation says that the satellite path is $2 \times 22\,500 - 3000 = 42\,000$ miles ($2 \times 36\,000 - 4800 = 67\,200$ km) longer than the terrestrial path and hence the extra delay is $42\,000/186\,000 = 67\,200/300\,000 = 0.225$ s. This represents $64\,000 \times 0.225 = 14\,400$ bits. This assumes that both cities are directly under the satellite; a more realistic assumption would be to assume that the cities were on the rim of the earth as seen from the satellite. Also the speed of propagation in an optical fibre is typically 0.7 of the free space speed.

2 Why is the path overhead not associated with the other overheads in the SDH frame structure?

The multiplex and regenerator section overheads support management functions on the line system. The path overhead supports management functions for the virtual container which stay with the virtual container as it moves from line system to line system.

3 Why could the physical interface described in Section 5.1 not simply be operated at 150 Mbit/s to provide a multipoint broadband interface?

The maximum length of the I.420 multipoint interface bus is determined by the time delays along the bus. This delay is such that a digit from a terminal at the far end of the bus is only displaced by a fraction of a digit period compared with a digit from a terminal at the NT1 end of the bus. If the bit rate is increased by a factor of about 1000 then the digit period would be reduced by the same factor and hence the allowable length of the bus would also decrease by the same factor, resulting in a maximum bus length of 0.2 metres!

ISDN Vocabulary

Here are some terms which are either newly created for ISDN or have specific meanings in an ISDN context. The French, German and Spanish equivalents are also provided.

access capability

F: *capacité d'accés*
G: *Anschlußkapazität*
S: *capacidad de acceso*

The number and type of the access channels at an ISDN access interface that are actually available for telecommunication purposes.

bearer service

F: *service support*
G: *Übermittlungsdienst*
S: *servicio portador*

A type of telecommunication service that provides the capability for the transmission of signals between user–network interfaces.

channel-associated signalling

F: *signalisation voie par voie*
G: *kanalgebundene Zeichengabe*
S: *señalización* asociada al canal

A model of signalling in which information relating to the traffic carried by a single channel is transmitted in the channel itself or in a signalling channel permanently associated with it.

common channel signalling

F: *signalisation sur voie commune, signalisation par canal sé*maphore
G: *Zentralkand–Zeichengabe*
S: *señalización por canal común*

A method of signalling in which signalling information relating to a multiplicity of circuits or functions or for network management, is conveyed over a single channel by addressed messages.

connection

F: *connexion, chaîne de connexion*
G: *Verbindung*
S: *conexión*

A concatenation of transmission channels or telecommunication circuits, switching and other functional units set up to provide for the transfer of signals between two or more points in a telecommunication network, to support a single communication.

connection attribute

F: *attribut de connexion*
G: *Verbindungsmerkmal*
S: *atributo de conexión*

A specified characteristic of an ISDN connection.

digital connection

F: *connexion numérique*
G: *digitale Verbindung*
S: *conexión digital*

A concatenation of digital transmission channels or digital telecommunication circuits, switching and other functional units set up to provide for the transfer of digital signals between two or more points in a telecommunication network, to support a single communication.

digital exchange

F: *commutateur numérique*
G: *Digitale Vermittlungstelle*
S: *central digital*

An exchange that switches digital signals by means of digial switching.

digital network, integrated digital network

F: *réseau numérique, réseau numérique intégré*
G: *Digitalnetz, integriertes Digitalnetz*
S: *red digital, red digital integrada*

A set of digital nodes and digital links that uses integrated transmission and switching to provide digital connections between two or more defined points to facilitate telecommunication between them.

integrated digital transmission and switching

F: *transmission et commutation numériques intégrées*
G: *integrierte digitale Übertragung und Vermittlung*
S: *transmisión y conmutación digitales integradas*

The direct (digital) concatenation of digital transmission and digital switching, that maintains a continuous digital transmission path.

integrated services digital network (ISDN)

F: *réseau numérique avec intégration des services* (RNIS)
G: *diensteintegrierendes* Digitalnetz (ISDN)
S: *red digital de servicios integrados* (RDSI)

An integrated services network that provides digital connection between user-network interfaces.

layer

F: *couche*
G: Schicht
S: *capa*

A conceptual region that embodies one or more functions between an upper and a lower logical boundary within a hierarchy of functions.

network termination

F: *terminaison de réseau*
G: Netzabschluß
S: *terminación de red*

Equipment that provides the functions necessary for the operation of the access protocols by the network.

non-switched connection

F: *connexion non commutée*
G: Festgeschaltete Verbindung
S: *conexión no conmutada*

A connection that is established without the use of switching, for example by means of hard-wired joints.

reference configuration

F: *configuration de référence*
G: Bezugskonfiguration
S: *configuración* de referencia

A combination of functional groups and reference points that shows possible network arrangements.

reference point

F: *point de référence*
G: Bezugspunkt
S: *punto de referencia*

A conceptual point at the conjunction of two non-overlapping functional groups.

service, telecommunication service

F: *service, service de télécommunications*
G: Dienst, *Fernmeldedienst*
S: *servicio, servicio de telecomunicación*

That which is offered by an Administration or Telecoms Provider to its customers in order to satisfy a specific telecommunication requirement.

service attribute, telecommunication service attribute

F: *attribut de service, attribut de service de télécommunications*
G: *Dienstmerkmal, Fernmeldedienstmerkmal*
S: *atributo de servicio, atributo de servicio de telecommunicación*

A specific characteristic of a telecommunication service.

switched connection

F: *connexion commutée*
G: *Vermittelte Verbindung*
S: *conexión conmutada*

A connection that is established by means of switching.

teleaction service [**telemetry service**]

F: *service de télé*action [*service de télemesure*]
G: *Fernwirk*–Dienst [*Telemetriedienst*]
S: *servicio de teleacción*

A type of telecommunication service that uses short messages, requiring a very low transmission rate, between the user and the network.

teleservice [**telecommunication service**]

F: *téléservice*
G: *Standardisierter Dienst* [*Teledienst*]
S: *teleservicio, servicio final*

A type of telecommunication service that provides the complete capability, including terminal equipment functions, for communication between users according to protocols established by agreement between Administrations and/or Telecoms Providers.

terminal equipment (**TE**)

F: *é*quipement terminal (ET)
G: *Endeinrichtung, Endgerät*
S: *equipo termiñal*

Equipment that provides the functions necessary for the operation of the access protocols by the user.

user access, user-network access

F: accès d'usager, accès d'usager-réseau
G: Anwenderzugang, Anwender-Netzzugang
S: acceso de usuario, acceso usuario-red

The means by which a user is connected to a telecommunication network in order to use the services and/or facilities of that network.

user-network interface

F: interface usager-réseau
G: Teilnehmer-Netz-Schnittstelle
S: interfaz usuario-red

The interface between the terminal equipment and a network termination at which interface the access protocols apply.

user-user protocol

F: protocole d'usager à usager
G: Anwender-Anwender-Protokoll
S: protocolo usuario-usuario

A protocol that is adopted between two or more users in order to ensure communication between them.

Index